SAXON MATH™

Intermediate 3, 4, 5

Teacher's Resource Handbook

Stephen Hake

SAXON™
A Harcourt Achieve Imprint

www.SaxonPublishers.com
1-800-284-7019

Printed in the U.S.A.

ISBN 978-1-602-77048-5

8 9 10 11 2266 20 19 18 17 16

4500621938 ^ B C D E F G

Table of Contents

Focus on Student Success

The Saxon Math program focuses on student success. The unique structure of Saxon Math instruction promotes student success through the sound educational practices of incremental development and continual review. Consider how most other mathematics programs are structured. Content is organized into topical chapters. A topic is developed rapidly over a period of a few weeks, sometimes too rapidly for students to grasp. Student exercises are characterized by practice that focuses on the day's lesson. When the chapter ends, the topic changes, and often practice of the topic ends as well. Students learn in blocks and likely forget in blocks. Thus, when standardized testing rolls around, many teachers feel an urgent need to review. Chapter organization might be good for reference, but it is not the best organization for learning. Incremental development and continual review are structural designs that improve student learning.

Incremental Development

With incremental development, topics are developed in small steps spread over time. One facet of a concept is taught and practiced before the next facet is introduced. Both facets are then practiced together until it is time for the third to be introduced. Instead of being organized in chapters that rapidly develop a topic and then move to the next strand, the content in the Saxon program is organized in a series of lessons that gradually develop concepts. This incremental approach provides students with time to solidify the prerequisite concepts and skills before they are introduced to the next step of instruction.

Continual Review

Through continual review, previously presented concepts are practiced frequently and extensively over the year. Saxon's cumulative daily practice strengthens students' grasp of concepts, improves their flexibility to work with several mathematical concepts at a time, and improves their long-term retention of concepts. Through incremental development of topics coupled with continual review, students are given the time to develop a deeper understanding of concepts and how to apply them.

Saxon Math Components

Core Program

Student Edition

Student Edition eBook
Complete student text on CD

Teacher's Manual
Two-volume spiral bound

Teacher's Manual eBook
Complete teacher edition on CD

Resources and Planner CD
An electronic pacing calendar with standards, plus assessment, reteaching, and instructional masters

Answer Key CD
Student answers for displaying to check homework

Meeting the Needs of All Students

Written Practice Workbook
No need to carry the textbook home

Power Up Workbook
Consumable worksheets for every lesson

Reteaching Masters
One for every lesson

Manipulatives Kit
Use with activities, investigations, examples, and Written Practice problems

Adaptations for Saxon Math
A complete, parallel program for special populations

Math Launch, Grade 3
A set of 11 thematic bulletin board posters and materials to accompany each of the 11 sections of Saxon Math Intermediate 3 textbook

English Learners Handbook
Provides guidance to the teacher for modifying instruction to accommodate English learners

Test-Taking Strategies Guide
Provides guidance for taking standardized and state tests

Building the Depth of the Standards

Instructional Masters

Activity masters, Test-Day Activities, Power-Up Worksheets, recording forms, and more

Performance Tasks

Performance Tasks with rubrics

Instructional Transparencies

Overhead transparencies of all instructional masters

Calculator Activities

Correlated to lessons

Solutions Manual

Solutions for all Student Edition problems

Instructional Posters

In English and Spanish

Online Activities

Real-world investigations and calculator and exploration activities

Instructional Presentations

Presentations for every lesson for instant teacher support

Tracking and Benchmarking the Standards

Assessment Guide

Assessments to check student progress, plus recording forms for easy tracking and analysis of scores

Test and Practice Generator CD

A tool that allows teachers to create unlimited test and practice items, in multiple formats and in both English and Spanish

Monitoring Student Progress eGradebook CD

Electronic gradebook to track progress on Cumulative and Benchmark Tests and to generate a variety of reports

Available in Spanish

Student Edition, Teacher's Manual, plus assessments, blackline masters, posters, test and practice generator questions, and online activities

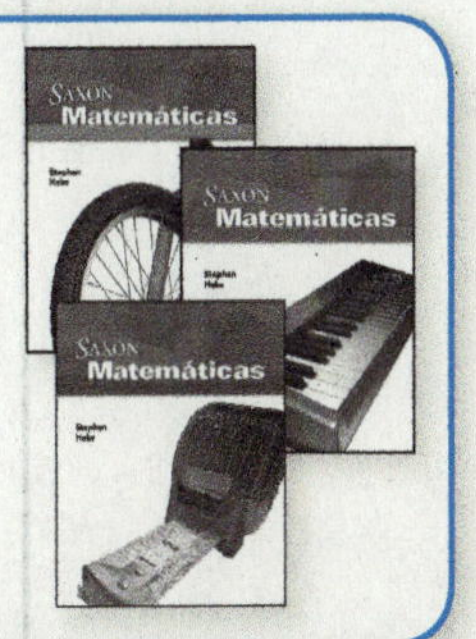

Beginning the Year

Before beginning the school year, new teachers should familiarize themselves with the Saxon program. Carefully read through the following sections of this booklet: *Teaching a Daily Lesson, Teaching an Investigation,* and *Assessing Student Progress.* These sections outline the daily implementation of Saxon Math. In these sections you will find various methods for using provided materials, such as answer forms and assessments. Consider how you might want to incorporate these methods into your classroom. This will help you to efficiently prepare for the school year.

Tip

The ***Saxon Math*** *Power-Up Workbook* that provides consumable worksheets for every lesson is available.

1. Photocopy and File Each Power Up.

The *Instructional Masters* booklet contains blackline masters for the Power-Up Worksheets. If your district offers photocopying services, you may wish to copy and file all the Power-Up Worksheets needed for the school year.

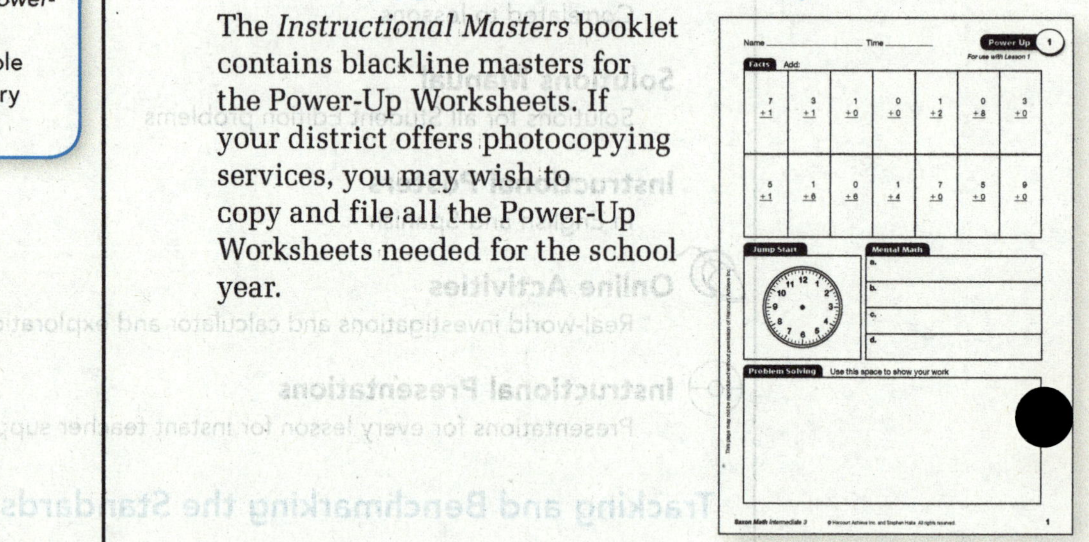

2. Photocopy Answer Forms.

Saxon Math encourages students to begin their assignments with the most difficult problems, so it is important for them to have a way of organizing their work. Blackline masters for the lesson recording forms are included in the *Instructional Masters* booklet. When using these forms, students can move quickly from problem to problem and stay focused when waiting for teacher assistance. These forms also make papers easy to check.

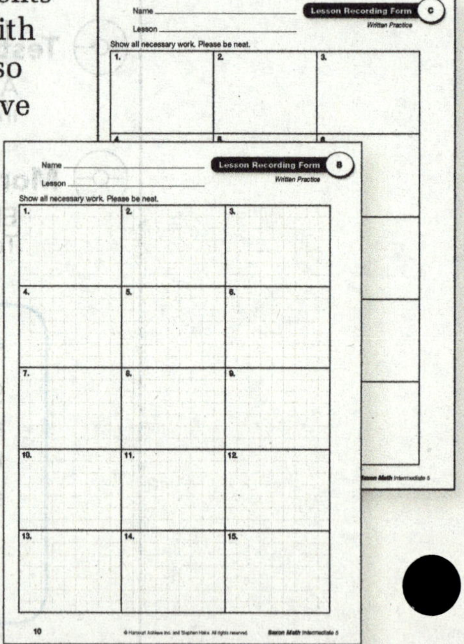

At the beginning of the year, decide which of the forms you would like your students to use. You may want to photocopy them in mass amounts and keep them in trays on a table where students can easily access them.

3. Photocopy Assessments.

The *Assessment Guide* contains blackline masters of two forms for each Cumulative Test.

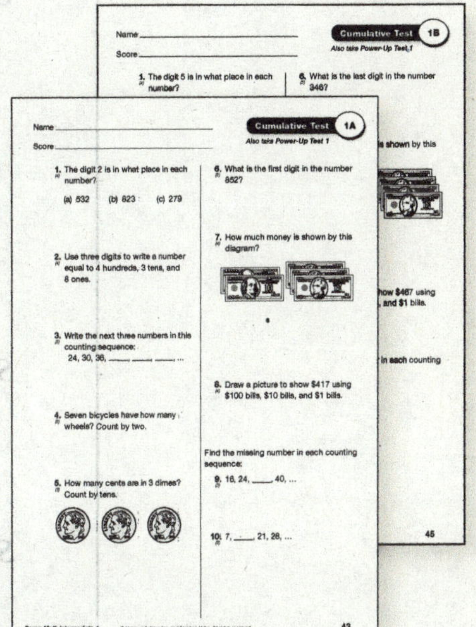
Name
Score
Cumulative Test 1B
Also take Power-Up Test 1

1. The digit 5 is in what place in each number?

6. What is the last digit in the number 346?

Name
Score
Cumulative Test 1A
Also take Power-Up Test 1

1. The digit 2 is in what place in each number?
(a) 532 (b) 823 (c) 279

2. Use three digits to write a number equal to 4 hundreds, 3 tens, and 8 ones.

3. Write the next three numbers in this counting sequence:
24, 30, 36, ___, ___, ___, ...

4. Seven bicycles have how many wheels? Count by two.

5. How many cents are in 3 dimes? Count by tens.

6. What is the first digit in the number 852?

7. How much money is shown by this diagram?

8. Draw a picture to show $417 using $100 bills, $10 bills, and $1 bills.

Find the missing number in each counting sequence:

9. 16, 24, ___, 40, ...

10. 7, ___, 21, 28, ...

Saxon Math Intermediate 4 © Harcourt Achieve Inc. and Stephen Hake. All rights reserved. 43

You may:

- Use one form as an original test and the other as a makeup test.
- Use both forms on test day to discourage copying.
- Use one form as an in-class practice.

At the beginning of the year, decide how you want to incorporate the two test forms into your classroom. You may want to photocopy the assessments as necessary, and place them in file folders and plastic crates for easy access.

4. Optional for Grades 4 and 5: Create Math Notebooks.

One way to help your students stay organized is to have each student create a math notebook. Students should bring their notebooks to math class every day.

To make a notebook, each student needs to bring the following materials:

- one-inch, three-ring binder
- spiral notebook
- three divider tabs
- three-ring supply pouch

Tip

For an inexpensive supply pouch, use a freezer bag. Just punch three holes in the bag, and place it in the front of the math notebook.

Steps for Making a Math Notebook

Step 1: Begin by placing the spiral notebook in the front of the binder. The spiral notebook should be used to take short, daily notes on the lesson.

Step 2: Next, place the three divider tabs behind the spiral notebook. Behind the first tab, place Student Progress Chart D: Facts Practice. This form can be found in the *Instructional Masters* booklet and is used by students to monitor their performance on the Power-Up Worksheets throughout the year. As the year progresses, students can place their completed Power-Up Worksheets behind the progress chart.

Step 3: Behind the second tab, have students place a sheet of paper to record their daily assignments. Students can then place their completed homework assignments behind the assignment sheet.

Step 4: If you are using the Written Practice answer forms, place extra copies behind the third divider. If answer forms are not being used, place notebook paper behind this divider.

Step 5: Finally, place the three-ring supply pouch in the front of the notebook. The pouch can hold students' pencils, rulers, protractors, and compasses. Because this is a math notebook, it should include only supplies needed during math class.

For students to succeed in the Saxon Math program, they must complete all of the math lessons each year. A pacing calendar—a plan that shows which lesson/investigation/test will occur each day of the school year—can help ensure that you complete the program.

Making a Pacing Calendar

A personalized schedule based on your school's calendar can be created using the Saxon Math Resources and Planner CD or you can use a copy of your school's calendar to create a pacing calendar by hand.

Use the following teaching schedule, to help you create a pacing calendar that will help you complete all of the lessons by the end of your school year. Beginning with the first day of school, assign a lesson for each day. On the tenth day of school, assign Lesson 10. On the eleventh school day, assign Test 1 and Performance Task 1. Then on the twelfth day, assign Investigation 1. Investigations occur after every tenth lesson, following the assessment day. After the first assessment, each assessment will occur after every fifth lesson. For example, after Lesson 15, you will give Test 2 and assign Test Day Activity 1. After Lesson 20, you will give Test 3 and assign Performance Task 2.

Creating a Pacing Calendar Using the Resources and Planner CD

When using the Saxon Math Resources and Planner CD, enter dates of school vacations, special events, and other days you will not be teaching a math lesson. The Resource and Planner CD then creates a schedule for the year. If unexpected school cancellations occur, you can easily use the Resource and Planner CD to produce a revised schedule.

Tip

Investigation 4 in *Intermediate 4* might require two days to complete. Be sure to account for this in your pacing calendar.

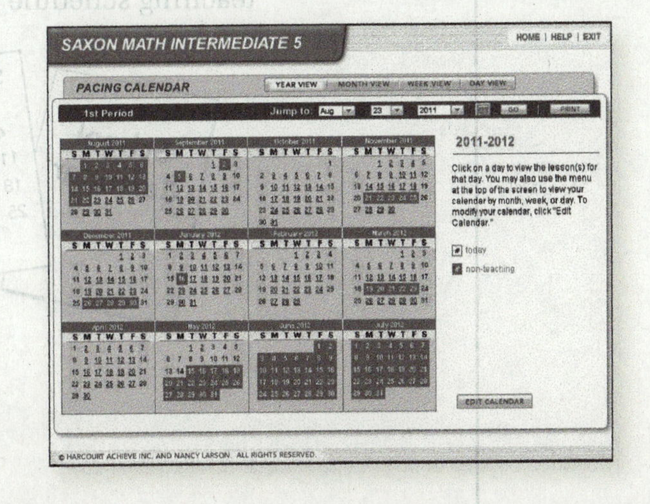

The program not only provides a pacing calendar yearly view but also a monthly view, a weekly view, and a daily view.

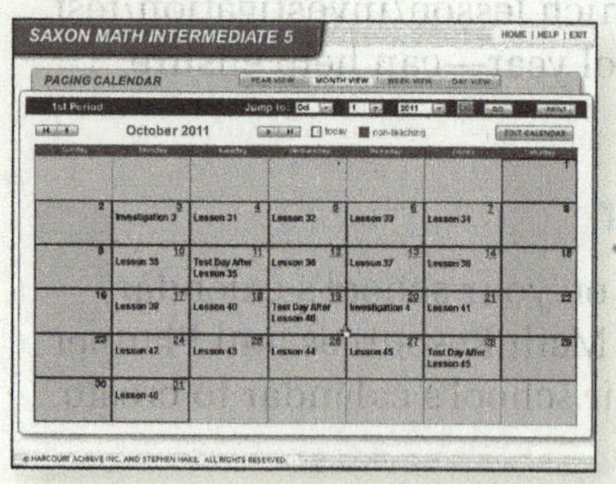

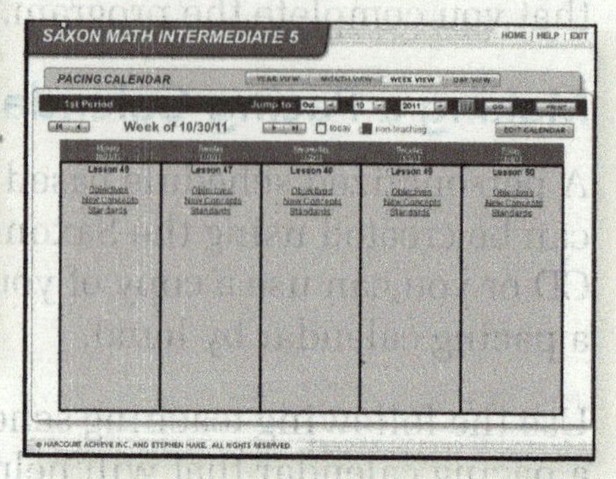

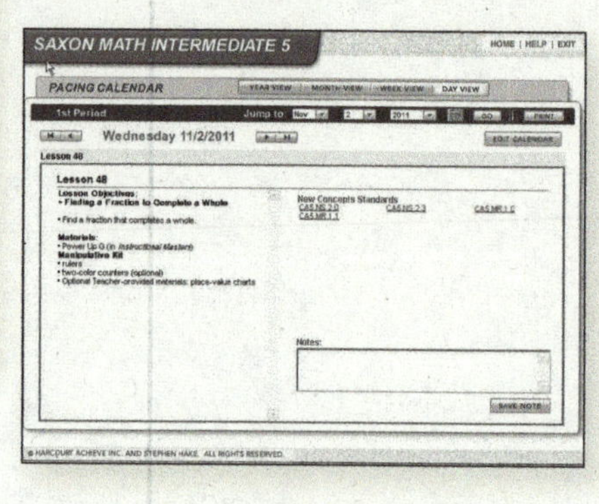

Creating a Pacing Calendar by Hand

To create a pacing calendar by hand, use the following steps:

Steps for Making a Pacing Calendar

Step 1: Get a copy of your school calendar, and enlarge it if possible.

Step 2: Cross out school holidays, vacation days, parent conference days, days dedicated to standardized testing, and any other days you know you will not be able to teach a math lesson.

Step 3: Count the number of days left for instruction (eliminating the last day of school).

Step 4: Write the number of the lesson you will teach each day above the school days on your calendar using the teaching schedule explained on page 7.

Pacing the Day

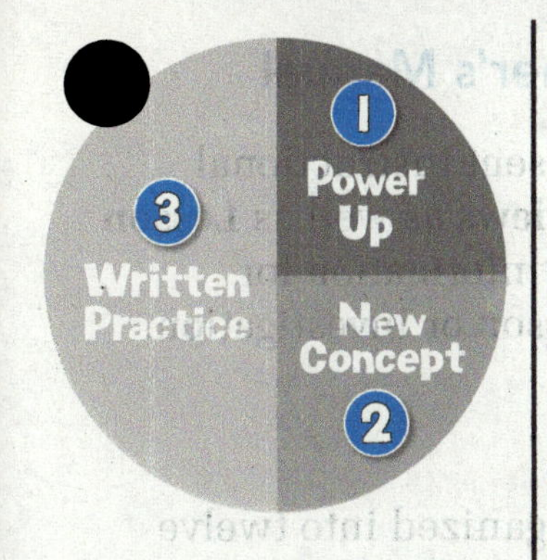

Saxon Math has a consistent lesson structure that enhances mastery of skills and concepts. Every lesson in Saxon Math follows the same three-part lesson plan. This regular format allows students to become comfortable with the lessons and to know what to expect each day.

The Saxon distributed approach is unique in that the focus of the day is mainly on the rich depth of content in the distributed Written Practice.

In a typical 60-minute class period, it is suggested that about half of the class period be used to have the students complete the Written Practice problems.

The chart below shows the approximate amount of time required to complete each section of a lesson. To minimize transition time, try to establish routines for moving from one activity to the next.

Daily Pacing Chart for 60 Minute Class Period

	Grade 3	Grade 4	Grade 5
Power Up			
Facts Practice	5 minutes	5 minutes	5 minutes
Jump Start	3 minutes		
Count Aloud		2 minutes	2 minutes
Mental Math	3 minutes	3 minutes	3 minutes
Problem Solving	4 minutes	5 minutes	5 minutes
Total	**15 minutes**	**15 minutes**	**15 minutes**
Lesson			
New Concept	10 minutes	10 minutes	10 minutes
Lesson Practice	5 minutes	5 minutes	5 minutes
Total	**15 minutes**	**15 minutes**	**15 minutes**
Written Practice			
Total	**Remainder of Class Time**	**Remainder of Class Time**	**Remainder of Class Time**

Planning Lessons Using the Teacher's Manual

The Teacher's Manual is organized to present instructional support in proper context. **Section Overviews** as well as **Lesson and Investigation Support Pages** provide information for teachers as they prepare to teach each lesson or investigation.

Section Overviews

The content of the Teacher's Manual is organized into twelve sections. Each section includes ten lessons and an investigation and is preceded by a Section Overview. The Section Overview consists of six pages and is a valuable source of information because it provides the teacher with information that will help with long-range planning. The title and a short description of each page of the Section Overview are given below.

- **Lesson Planner**
  Gives the new concepts being presented as well as the materials and resources needed for that section.

- **Math Highlights**
  Gives a list of big ideas for the section, along with key questions to ask to assess understanding. Identifies math content and process skills that will be covered.

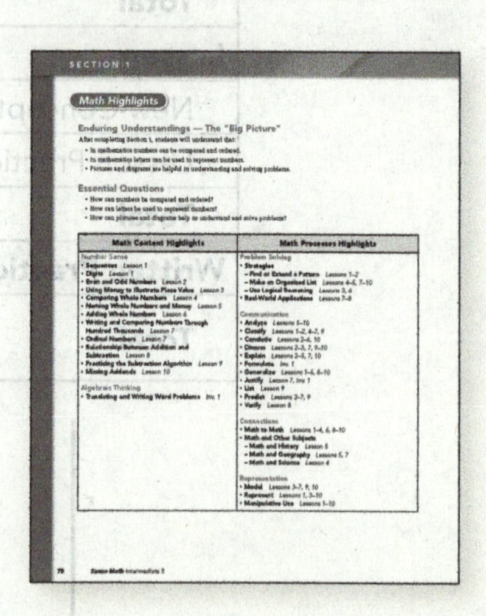

SECTION 1

Math Highlights

Enduring Understandings — The "Big Picture"

After completing Section 1, students will understand that:

- In mathematics numbers can be compared and ordered.
- In mathematics letters can be used to represent numbers.
- Pictures and diagrams are helpful in understanding and solving problems.

Essential Questions

- How can numbers be compared and ordered?
- How can letters be used to represent numbers?
- How can pictures and diagrams help us understand and solve problems?

Math Content Highlights	Math Processes Highlights
Number Sense • Sequences Lesson 1 • Digits Lesson 1 • Even and Odd Numbers Lesson 2 • Using Money to Illustrate Place Value Lesson 3 • Comparing Whole Numbers Lesson 4 • Naming Whole Numbers and Money Lesson 5 • Adding Whole Numbers Lesson 6 • Writing and Comparing Numbers Through Hundred Thousands Lesson 7 • Ordinal Numbers Lesson 7 • Relationship Between Addition and Subtraction Lesson 8 • Practicing the Subtraction Algorithm Lesson 9 • Missing Addends Lesson 10 Algebraic Thinking • Translating and Writing Word Problems Inv. 1	Problem Solving • Strategies – Find or Extend a Pattern Lessons 1–2 – Make an Organized List Lessons 4–5, 7–10 – Use Logical Reasoning Lessons 3, 6 • Real-World Applications Lessons 7–8 Communication • Analyze Lessons 1–10 • Classify Lessons 1–2, 4–7, 9 • Conclude Lessons 2–6, 10 • Discuss Lessons 2–3, 7, 9–10 • Explain Lessons 2–5, 7, 10 • Formulate Inv. 1 • Generalize Lessons 1–6, 8–10 • Justify Lesson 7, Inv. 1 • List Lesson 9 • Predict Lessons 2–7, 9 • Verify Lesson 8 Connections • Math to Math Lessons 1–4, 6, 8–10 • Math and Other Subjects – Math and History Lesson 5 – Math and Geography Lessons 5, 7 – Math and Science Lesson 4 Representation • Model Lessons 3–7, 9, 10 • Represent Lessons 1, 3–10 • Manipulative Use Lessons 1–10

7B Saxon Math Intermediate 3

- **Differentiated Instruction**
  Directs teachers to features in the Teacher's Manual that can be used to customize instruction. Also provided is a list of additional resources that are available to support differentiated instruction for lessons presented in the section as well as a list of technology resources available for both student and teacher.

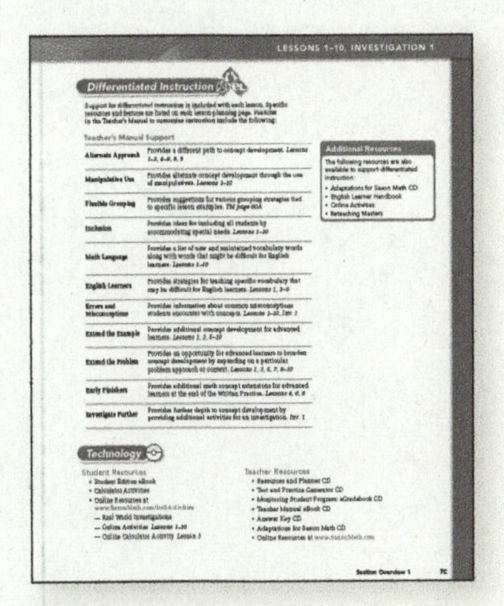

- **Cumulative Assessment**
  Provides the Power-Up Tests, Cumulative Tests, and Performance Task that will be administered for that section. The Evidence of Learning at the bottom of this page identifies the concepts or skills that students should be able to demonstrate by the end of the section of lessons.

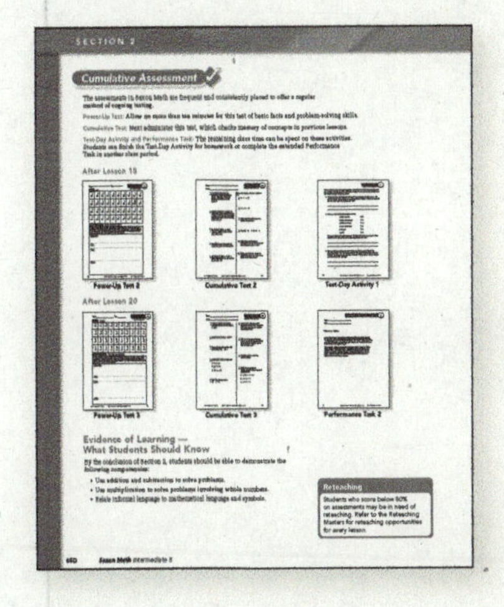

- **Benchmarking and Tracking Standards**
  Provides descriptions of the program resources that benchmark and track student progress, as well as instructions on how to generate a report that will determine which standards were assessed and the level of mastery for each student.

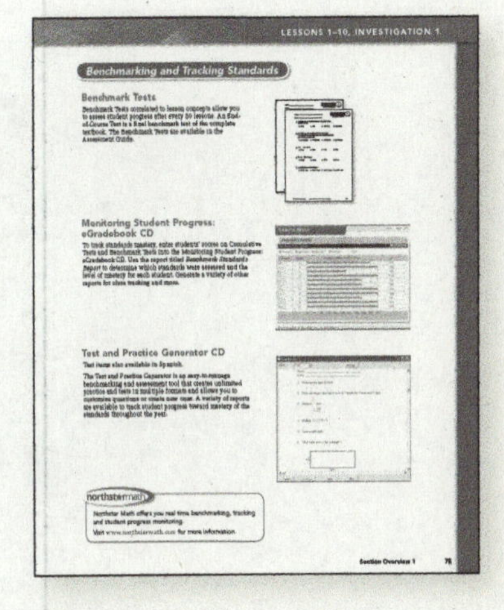

- **Content Trace**
  Illustrates the distributed approach of Saxon Math by showing lessons in which new concepts for the section are taught, practiced, and assessed. The concept presented in most lessons is a building block for future lessons. The *Looking Forward* section located on the last page of each lesson and investigation in the Teacher's Manual shows where that concept will be further developed.

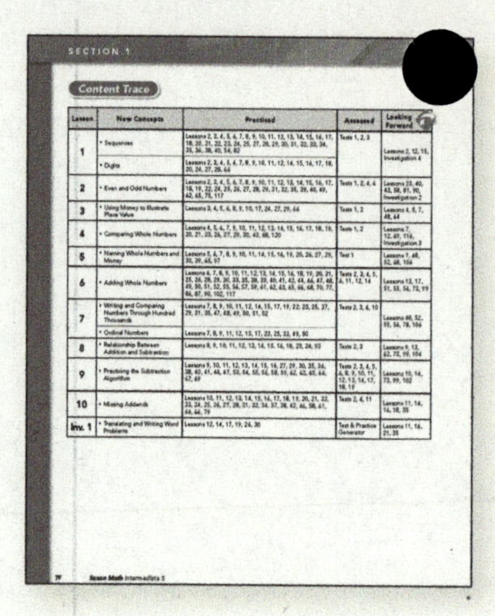

SECTION 1

Content Trace

Lesson	New Concepts	Practiced	Assessed	Looking Forward
1	• Sequences	Lessons 2, 3, 4, 5, 6, 7, 8, 9, 10, 11, 12, 13, 14, 15, 16, 17, 18, 20, 21, 22, 23, 24, 25, 27, 28, 29, 30, 31, 32, 33, 34, 35, 36, 38, 40, 54, 82	Tests 1, 2, 3	Lessons 2, 12, 15, Investigation 4
	• Digits	Lessons 2, 3, 4, 5, 6, 7, 8, 9, 10, 11, 12, 14, 15, 16, 17, 18, 20, 24, 27, 28, 66		
2	• Even and Odd Numbers	Lessons 2, 3, 4, 5, 6, 7, 8, 9, 10, 11, 12, 13, 14, 15, 16, 17, 18, 19, 22, 24, 25, 26, 27, 28, 29, 31, 32, 35, 39, 40, 49, 62, 65, 75, 117	Tests 1, 2, 4, 6	Lessons 23, 40, 43, 58, 81, 90, Investigation 2
3	• Using Money to Illustrate Place Value	Lessons 3, 4, 5, 6, 8, 9, 10, 17, 24, 27, 29, 66	Tests 1, 2	Lessons 4, 5, 7, 48, 64
4	• Comparing Whole Numbers	Lessons 4, 5, 6, 7, 9, 10, 11, 12, 13, 14, 15, 16, 17, 18, 19, 20, 21, 23, 25, 27, 29, 30, 43, 68, 120	Tests 1, 2	Lessons 7, 12, 69, 116, Investigation 3
5	• Naming Whole Numbers and Money	Lessons 5, 6, 7, 8, 9, 10, 11, 14, 15, 16, 19, 20, 26, 27, 29, 30, 39, 65, 97	Test 1	Lessons 7, 48, 52, 68, 104
6	• Adding Whole Numbers	Lessons 6, 7, 8, 9, 10, 11, 12, 13, 14, 15, 16, 18, 19, 20, 21, 25, 26, 28, 29, 30, 33, 35, 38, 39, 40, 41, 42, 44, 46, 47, 48, 49, 50, 51, 52, 55, 56, 57, 59, 61, 62, 63, 65, 66, 68, 70, 77, 86, 87, 90, 102, 117	Tests 2, 3, 4, 5, 6, 11, 12, 14	Lessons 13, 17, 51, 55, 56, 73, 99
7	• Writing and Comparing Numbers Through Hundred Thousands	Lessons 7, 8, 9, 10, 11, 12, 14, 15, 17, 19, 22, 23, 25, 27, 29, 31, 35, 47, 48, 49, 50, 51, 52	Tests 2, 3, 4, 10	Lessons 48, 52, 55, 56, 78, 106
	• Ordinal Numbers	Lessons 7, 8, 9, 11, 12, 15, 17, 23, 25, 33, 49, 50		
8	• Relationship Between Addition and Subtraction	Lessons 8, 9, 10, 11, 12, 13, 14, 15, 16, 19, 23, 24, 93	Tests 2, 3	Lessons 9, 13, 62, 72, 99, 104
9	• Practicing the Subtraction Algorithm	Lessons 9, 10, 11, 12, 13, 14, 15, 16, 27, 29, 30, 35, 36, 38, 40, 41, 44, 47, 53, 54, 55, 56, 58, 59, 62, 63, 65, 66, 67, 69	Tests 2, 3, 4, 5, 6, 8, 9, 10, 11, 12, 13, 14, 17, 18, 19	Lessons 10, 14, 73, 99, 102
10	• Missing Addends	Lessons 10, 11, 12, 13, 14, 15, 16, 17, 18, 19, 20, 21, 22, 23, 24, 25, 26, 27, 28, 31, 32, 34, 37, 38, 42, 46, 58, 61, 64, 66, 79	Tests 2, 4, 11	Lessons 11, 14, 16, 18, 35
Inv. 1	• Translating and Writing Word Problems	Lessons 12, 14, 17, 19, 26, 30	Test & Practice Generator	Lessons 11, 16, 21, 35

Lesson Support Pages

In addition to the Section Overview, each lesson and investigation is preceded by two support pages. These pages will help teachers with daily lesson preparation. The title and a description of each suppport page is given below.

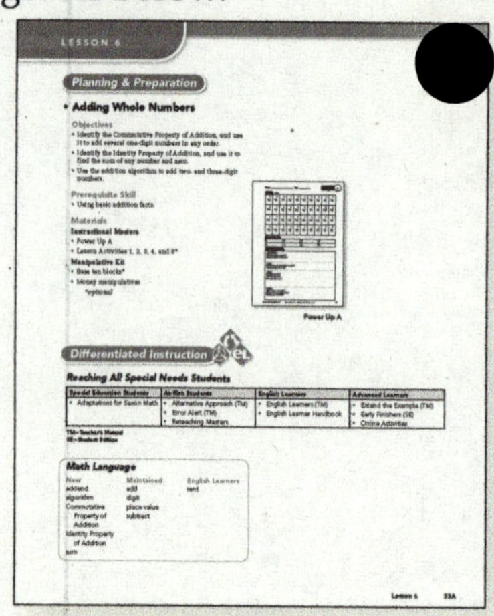

- **Planning and Preparation**
  Provides a list of Objectives that students will learn in the lesson and a list of Prerequisite Skills that are necessary for understanding the concept presented in the lesson. This page also includes a Materials list that provides materials included with the program as well as materials that the teacher will need to provide. The Differentiated Instruction section directs teachers to resources available for that lesson that can be used to customize instruction for all special needs students. The Math Language box provides glossary terms used in the lesson and alerts teachers to terms that may be difficult for English learners.

- **Problem Solving Discussion**
  Follows the four-step problem solving plan and guides the discussion as the class works together to complete the problem solving activity at the end of each Power Up using the strategies given. When possible have students show how additional strategies can be used to solve the problem.

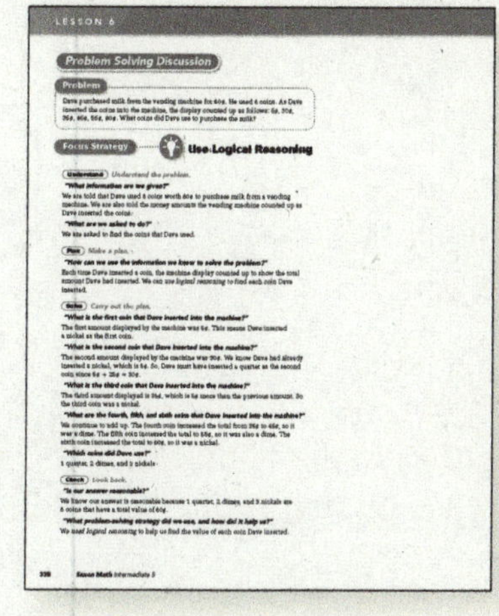

Investigation Support Pages

- **Flexible Grouping**
  Includes ideas for various grouping strategies tied to specific lessons. These flexible grouping ideas give students the opportunity to work with other students in an interactive and encouraging way.

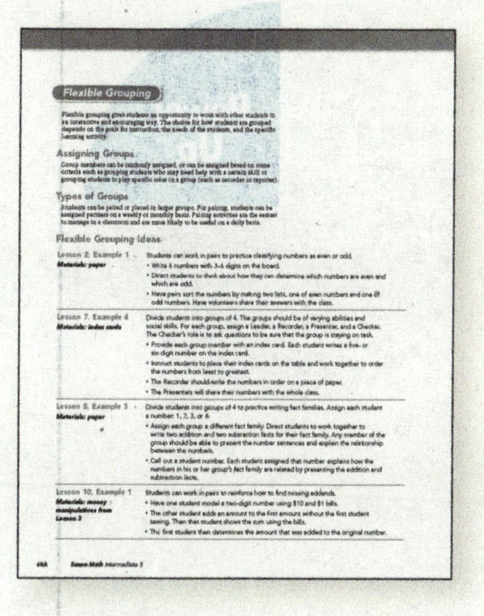

- **Planning and Preparation**
  Provides objectives for the investigation and a list of prerequisite skills necessary for the investigation. This page also includes a materials list, a vocabulary list, and a list of resources that will help teachers differentiate instruction to reach all special needs students.

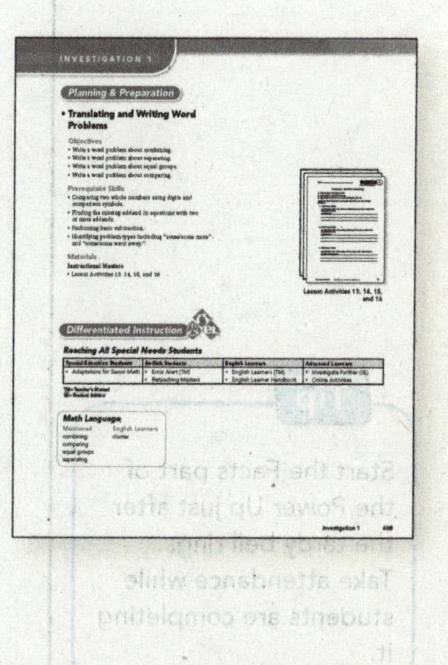

Teaching a Daily Lesson

Effective implementation is critical if students are to achieve optimal benefits using ***Saxon Math*** *Intermediate 3–5*. This section describes practices that have proven effective in classrooms over many years. Practice has proven that students learn mathematics best by working on problems themselves. Therefore, the focus of class time should be on providing the maximum amount of time for students to work productively on the prescribed problems.

> **REMEMBER!** Daily lessons must be taught in sequence. Cover one lesson, investigation, or assessment each day.

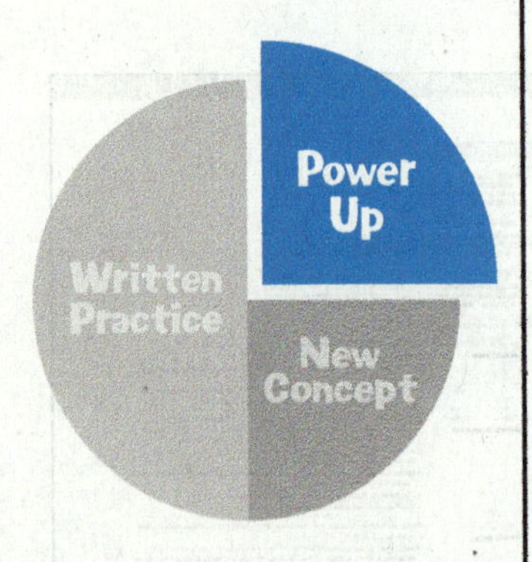

Power Up

In ***Saxon Math*** *Intermediate 3–5* students' prior knowledge is activated and built upon during the Power Up. The Power Up at the beginning of every lesson provides daily reinforcement and building of:

- basic skills and concepts
- mental math
- problem-solving strategies

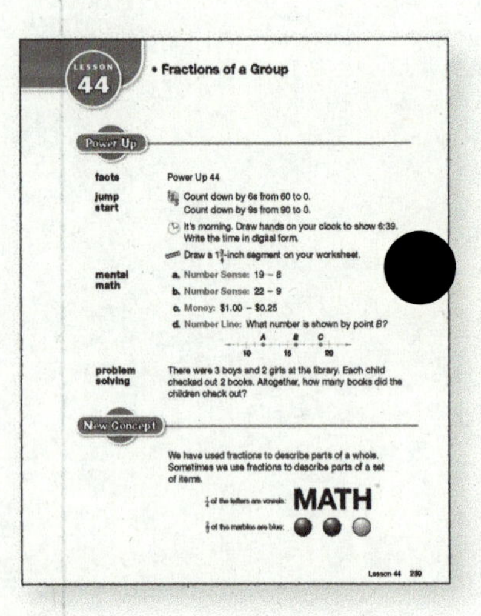
LESSON 44

• Fractions of a Group

Power Up

facts — Power Up 44

jump start — Count down by 6s from 60 to 0. Count down by 9s from 90 to 0.

It's morning. Draw hands on your clock to show 6:39. Write the time in digital form.

Draw a $1\frac{3}{4}$-inch segment on your worksheet.

mental math —
a. Number Sense: 19 − 8
b. Number Sense: 22 − 9
c. Money: $1.00 − $0.25
d. Number Line: What number is shown by point *B*?

problem solving — There were 3 boys and 2 girls at the library. Each child checked out 2 books. Altogether, how many books did the children check out?

New Concept

We have used fractions to describe parts of a whole. Sometimes we use fractions to describe parts of a set of items.

$\frac{1}{4}$ of the letters are vowels: MATH

$\frac{2}{3}$ of the marbles are blue.

Lesson 44 239

Daily work on these problems results in automaticity of basic skills and mastery of mental math and problem-solving strategies. For those students who need extra time, the Saxon approach allows for mastery gradually over time.

It is important that a **Facts Practice** precedes each lesson. It is critical that students achieve automaticity of number facts. Fluency will allow students to compute mentally, to estimate answers, and to enjoy success in later math courses.

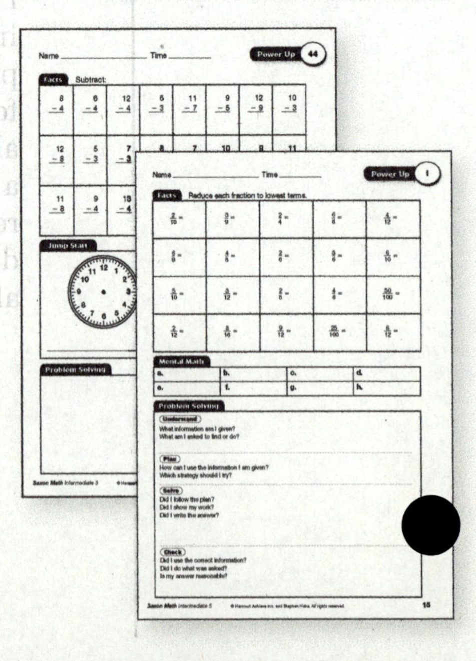
Name ______ Time ______ Power Up 44

Facts Subtract.

Jump Start

Problem Solving

Name ______ Time ______ Power Up I

Facts Reduce each fraction to lowest terms.

Mental Math

Problem Solving

Understand — What information am I given? What am I asked to find or do?

Plan — How can I use the information I am given? Which strategy should I try?

Solve — Did I follow the plan? Did I show my work? Did I write the answer?

Check — Did I use the correct information? Did I do what was asked? Is my answer reasonable?

> **Tip**
>
> Start the Facts part of the Power Up just after the tardy bell rings. Take attendance while students are completing it.

Overhead timers can be used to measure students' Facts Practice completion times. Most timers have the option of "counting up." This allows students to look up at the timer to see how long it took them to finish the Facts Practice.

Tip

Give grades for Facts Practices only on Cumulative Test days.

Students should begin the Facts section of the Power Up on your signal. Limit the time for each Facts section to three minutes or less. Timing provides motivation and encourages students to strive to improve on their previous scores.

Answers for the Facts section of the Power Ups are found in the *Teacher Manuals* and the *Instructional Transparencies.* After time is up, quickly call out the answers or display the answer transparency of the Power Up for self-checking. Students should correct errors and complete the test as part of the day's assignment if they are unable to finish within the allotted time.

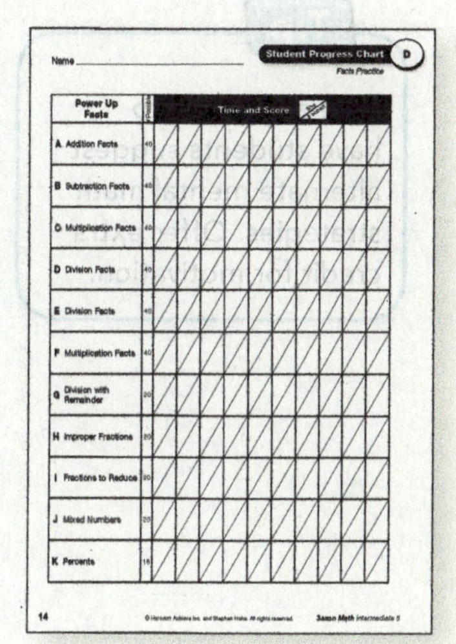

Name ____________

Student Progress Chart D

Facts Practice

Power Up Facts	# Possible	Time and Score
A Addition Facts	40	
B Subtraction Facts	40	
C Multiplication Facts	40	
D Division Facts	40	
E Division Facts	40	
F Multiplication Facts	40	
G Division with Remainder	20	
H Improper Fractions	20	
I Fractions to Reduce	20	
J Mixed Numbers	20	
K Percents	15	

14 *Saxon Math Intermediate 5*

Students can write their individual times and scores on their worksheet and Grade 4 and 5 students can later transfer the data to their Student Progress Chart in their notebooks.

REMEMBER! The time invested in Facts Practice is repaid in your students' ability to work more quickly.

Follow the facts practice with the **Jump Start** activity in grade 3 and with the **Count Aloud** activity in grades 4 and 5.

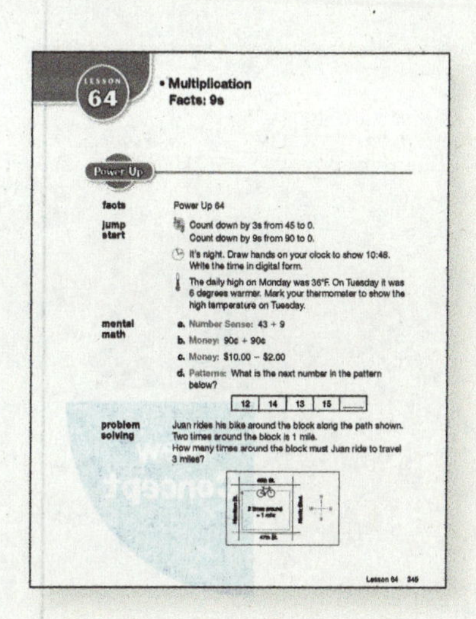

LESSON 64

• Multiplication Facts: 9s

Power Up

facts — Power Up 64

jump start — Count down by 3s from 45 to 0. Count down by 9s from 90 to 0.

It's night. Draw hands on your clock to show 10:48. Write the time in digital form.

The daily high on Monday was 36°F. On Tuesday it was 6 degrees warmer. Mark your thermometer to show the high temperature on Tuesday.

mental math

a. Number Sense: 43 + 9

b. Money: 90¢ + 90¢

c. Money: \$10.00 − \$2.00

d. Patterns: What is the next number in the pattern below?

12	14	13	15	___

problem solving — Juan rides his bike around the block along the path shown. Two times around the block is 1 mile. How many times around the block must Juan ride to travel 3 miles?

Lesson 64 345

The **Jump Start** activity in grade 3 provides practice for numeration and measurement skills and concepts. The Jump Start activity has students practice numeration skills such as:

- skip counting
- writing a number in standard or expanded form
- ordering a set of numbers from least to greatest

Students also practice measurement skills by:

- drawing hands on a clock
- marking a thermometer to show a certain temperature
- using a ruler to draw a line segment a particular length

The **Count Aloud** activity in grades 4 and 5 provides practice [illegible] numeration skills and concepts such as skip counting. As a cl[illegible] activity, students are asked to skip count by whole numbers and fractions.

Next, complete the **Mental Math** activity. At the beginning of the year, read the problems aloud while students follow along in the textbook. Later, read the problems aloud while students' books are closed. Students should perform calculations mentally and use their Power-Up Worksheet for recording their answers (not for calculation). Go over the answers with your students. Encourage them to explain the strategies they used to solve the problems.

Tip

Take a moment to have students suggest alternate mental math strategies. Offer extra credit for motivation.

Following the Mental Math section is a **Problem Solving** activity. These activities are designed to reinforce skills needed to learn more advanced mathematical concepts. They provide students with the opportunity to develop and practice problem solving strategies and to communicate ideas with one another. The problem should be tackled as a whole class activity, providing a problem solving experience for all students without intimidating students who are more hesitant.

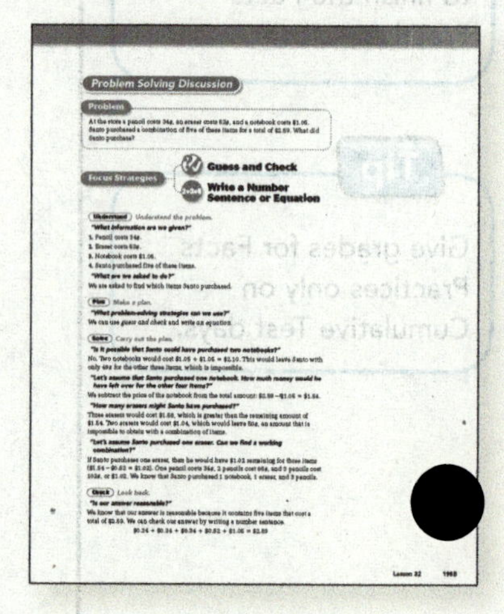
Problem Solving Discussion

Problem

At the store a pencil costs 34¢, an eraser costs 52¢, and a notebook costs $1.05. Santo purchased a combination of five of these items for a total of $2.59. What did Santo purchase?

Focus Strategies: Guess and Check; Write a Number Sentence or Equation

Understand Understand the problem.

"What information are we given?"

1. Pencil costs 34¢.
2. Eraser costs 52¢.
3. Notebook costs $1.05.
4. Santo purchased five of these items.

"What are we asked to do?"

We are asked to find which items Santo purchased.

Plan Make a plan.

"What problem-solving strategies can we use?"

We can use *guess and check* and write an equation.

Solve Carry out the plan.

"Is it possible that Santo could have purchased two notebooks?"

No. Two notebooks would cost $1.05 + $1.05 = $2.10. This would leave Santo with only 49¢ for the other three items, which is impossible.

"Let's assume that Santo purchased one notebook. How much money would he have left over for the other four items?"

We subtract the price of the notebook from the total amount: $2.59 − $1.05 = $1.54.

"How many erasers might Santo have purchased?"

Three erasers would cost $1.56, which is greater than the remaining amount of $1.54. Two erasers would cost $1.04, which would leave 50¢, an amount that is impossible to obtain with a combination of items.

"Let's assume Santo purchased one eraser. Can we find a working combination?"

If Santo purchases one eraser, then he would have $1.02 remaining for three items ($1.54 − $0.52 = $1.02). One pencil costs 34¢, 2 pencils cost 68¢, and 3 pencils cost 102¢, or $1.02. We know that Santo purchased 1 notebook, 1 eraser, and 3 pencils.

Check Look back.

"Is our answer reasonable?"

We know that our answer is reasonable because it contains five items that cost a total of $2.59. We can check our answer by writing a number sentence.

$0.34 + $0.34 + $0.34 + $0.52 + $1.05 = $2.59

Lesson 32 195B

A discussion guide that follows the four-step problem-solving plan is provided in the Teacher's Manual, but you are encouraged to solicit problem solving strategies from your students. Students can use their Power Up worksheet for sketches and calculations during this activity.

▶ *The entire Power Up should take less than fifteen minutes to complete.*

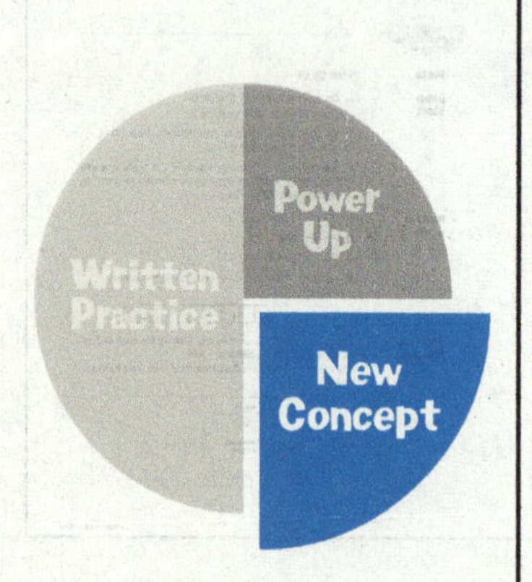

New Concept

Following the Power Up, present the New Concept. A New Concept will be presented each day; however, the student will only be required to learn a small part of the concept that day. Students will be building on the concept throughout the year. The presentation of the New Concept should be brief to maximize the time students have to solve problems in class.

Tip

Do not have students take notes on the entire lesson. This could take up too much class time. Have students take notes on the new lesson vocabulary and some of the example problems.

It is important to read over the lesson the night before presenting it to the class. Do not read the lesson word for word to your class. Use the instructional notes provided in your Teacher's Manual to enhance lesson instruction. The instructional notes will provide:

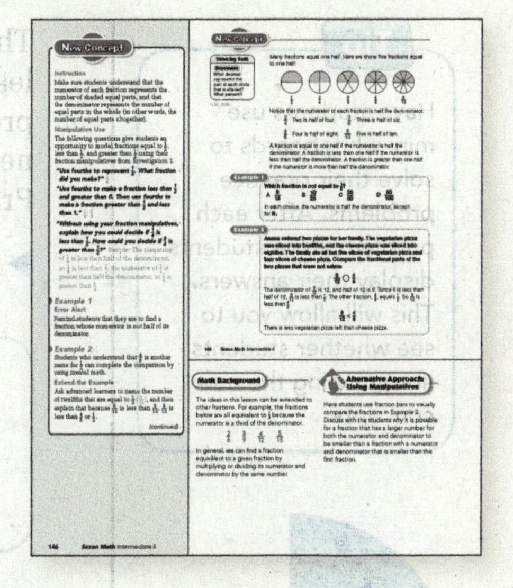

- Suggestions for ensuring student understanding of lesson material.
- Questions to ask students during lesson development.
- Additional activities, suggestions, and references to aid in the instruction of students with special needs or those struggling in mathematics.

The Teacher's Manual also provides suggestions for manipulative use within lessons and investigations. In the Written Practice, the Manipulative Use feature calls attention to problems that require students to use measurement tools such as rulers, protractors, and compasses.

Example problems are provided in each lesson to build depth and expand student knowledge of the New Concept. Model the example problems contained in the lesson, but resist the urge to show more examples. Students will see plenty of examples in the Lesson Practice and, over time, in the Written Practices.

To help students understand how and why the math presented in the New Concept works, the following items can be found in the Student Edition:

- Thinking Skill questions
- Reading Math hints
- Math Language tips

Some New Concepts will also have a hands-on activity. The activity will allow students to work together and to use manipulatives to explore the new concept.

▶ *Strive to keep the New Concept presentation brief.*

> **Tip**
>
> Have students use mini whiteboards to solve their practice problems. After each problem, have students display their answers. This will allow you to see whether students are grasping the new concepts.

The **Lesson Practice** requires students to apply the skills they learned in the New Concept. Guide students as they work the problems, and provide immediate feedback to students who need it. Students should always solve all problems in the Lesson Practice.

▶ *The Practice can usually be completed in five minutes.*

> **REMEMBER!** Mastery of the concept is not expected on the initial day of instruction. Work toward conceptual understanding, and guide the practice of the New Concept during the next few Written Practices.

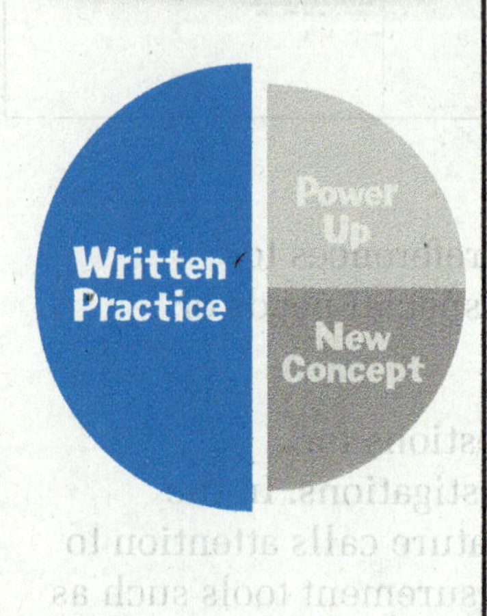

Written Practice

The Written Practice is the most important section of the daily lesson. It provides independent practice on the New Concept, as well as previously taught concepts and skills, ensuring that students gain and retain math skills. Once a skill has been taught, students are asked to use higher-order **thinking skills** and applications of that concept. The distributed mixed practice is unpredictable and, therefore, challenging for students.

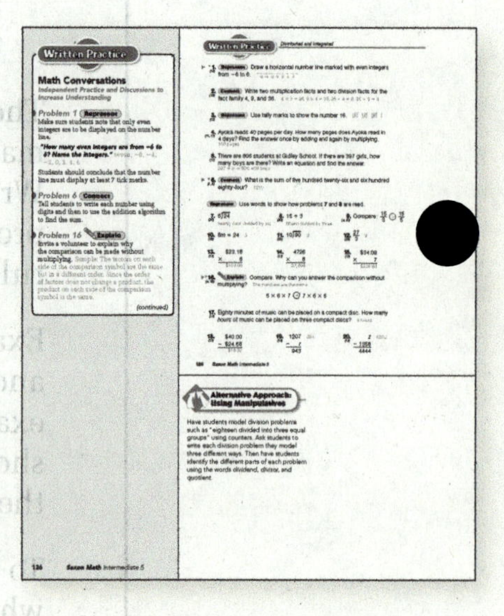

Written Practice

Math Conversations
Independent Practice and Discussions to Increase Understanding

Problem 1 (Represent)
Make sure students note that only even integers are to be displayed on the number line.
"How many even integers are from −6 to 6? Name the integers."
Students should conclude that the number line must display at least 7 tick marks.

Problem 6 (Connect)
Tell students to write each number using digits and then to use the addition algorithm to find the sum.

Problem 16 (Explain)
Invite a volunteer to explain why the comparison can be made without **multiplying.** Sample: The factors on each side of the comparison symbol are the same but in a different order. Since the order of factors does not change a product, the product on each side of the comparison symbol is the same.

(continued)

126 *Saxon Math* Intermediate 5

Written Practice *Distributed and Integrated*

1. (Represent) Draw a horizontal number line marked with even integers from −6 to 6.
2. (Connect) Write two multiplication facts and two division facts for the fact family 4, 9, and 36.
3. (Represent) Use tally marks to show the number 16.
4. Ayoka reads 40 pages per day. How many pages does Ayoka read in 4 days? Find the answer once by adding and again by multiplying.
5. There are 806 students at Gidley School. If there are 397 girls, how many boys are there? Write an equation and find the answer.
6. (Connect) What is the sum of five hundred twenty-six and six hundred eighty-four?

(Represent) Use words to show how problems 7 and 8 are read.

16. (Explain) Compare. Why can you answer the comparison without multiplying?

$5 \times 6 \times 7 \bigcirc 7 \times 6 \times 5$

17. Eighty minutes of music can be placed on a compact disc. How many hours of music can be placed on three compact discs?

Alternative Approach: Using Manipulatives

Have students model division problems such as "eighteen divided into three equal groups" using counters. Ask students to write each division problem they model three different ways. Then have students identify the different parts of each problem using the words *dividend, divisor,* and *quotient.*

During the Written Practice teachers should monitor students' work to ensure that they are correctly applying what they have learned. Students should always solve all problems in the Written Practice section. The problems not finished in class, can be completed as homework.

Tip

Lesson reference numbers appear below each problem number to indicate the lesson in which the concept was taught.

Tip

If you have students work on the most difficult problems first, they will be able to receive instant feedback about more-advanced concepts.

Tip

Students who finish the Written Practice problems before the end of the period may work on an Early Finishers problem.

There are several methods for completing the Written Practice. Many successful teachers have adapted the following procedure to suit the needs of their classroom:

Step 1: As you prepare for each lesson, notice that several problems have a star beside them. Students may find the problems with stars the most difficult. Many times these problems cover recently presented concepts or concepts that many students missed on the most recent assessment.

Step 2: After completing the Lesson Practice, students should take out a Written Practice answer form. If students need additional forms, they can retrieve them from the tray of extra forms copied at the beginning of the year. The Written Practice answer forms are located in the *Instructional Masters* booklet.

Step 3: Students should begin the Written Practice by first completing problems with stars.

Step 4: As students work on the starred problems, walk around the room and monitor students' work. Alternately, you may prefer to complete these problems as a guided class activity.

Step 5: When students have correctly finished the starred problems, they may begin working on the remainder of the Written Practice.

Step 6: Continue monitoring students as they work on the Written Practice.

Step 7: Problems not finished during the class period become homework for students to complete before the next math class.

We recommend **correcting homework** from the previous day in class immediately before beginning the Written Practice for the new lesson. Following the steps for completing the Written Practice should leave fewer homework problems to correct the next day.

Tip

You may assist students who need help with individual problems on homework after the other students begin work on the new Written Practice.

Correct the Written Practice by calling out or displaying the answers. The Answer Key CD provides answers to display to check homework. For optimal results, students should correct missed problems before working on the new Written Practice. Although you should review problems that presented difficulty for the majority of students, you should not spend excessive time on the previous day's homework. Instead, you can remedy identified problem areas by demonstrating similar problems as they occur on upcoming Written Practices. For more individualized remediation, see the *Providing Remediation* section of this booklet.

Tip

The ***Saxon Math*** *Solutions Manual* provides solutions for all Written Practice problems.

The Written Practice section is designed for practice, not for grading. We believe that a student's grade should represent his or her ability to do mathematics, and the assessments are the best indicator of that ability. This is not to imply that tests are more important than daily practice. In fact, without daily practice students will not be prepared to perform well on tests. Since the purpose of the Written Practice is practice rather than assessment, different grading strategies should be used.

Students are encouraged to seek help with daily assignments whenever they need it, so the grading strategy should simply reflect students' diligence in completing the work. Students are generally required to complete one Written Practice assignment per day. Each assignment should be due the next non-test day after it is assigned.

▶ *Allow at least half of your class time for Written Practice.*

Teaching an Investigation

Tip

Overhead transparencies for all Lesson and Investigation Activities are included with the *Instructional Transparencies.*

A teacher-directed investigation—an in-depth look at a particular concept—follows every tenth lesson. Investigations will:

- activate prior knowledge
- use teacher models
- provide guided practice
- ensure teacher-student interaction

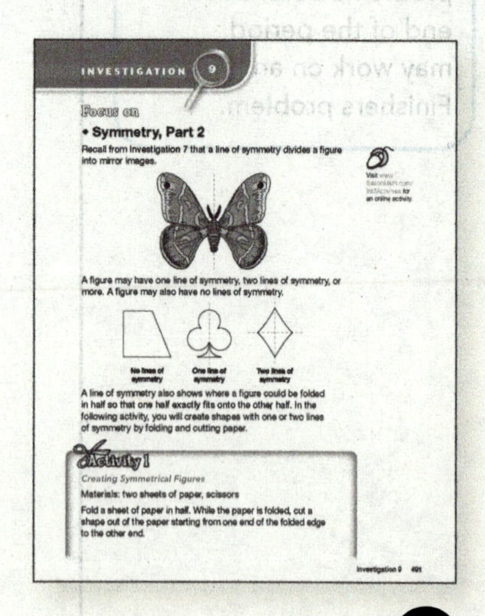

Investigations usually require a full class period to complete and are designed to be whole-class activities. Some investigations contain activities that use the manipulative approach to learning. Investigations may call for the use of one or more Lesson Activity Masters. These can be found in the *Instructional Masters* booklet.

The Investigate Further section at the end of some investigations provides an opportunity for students to expand their knowledge of the investigation concepts, use higher-order thinking skills, and explore more connections. This section can be completed as a class activity or it can be used as an assignment for advanced learners.

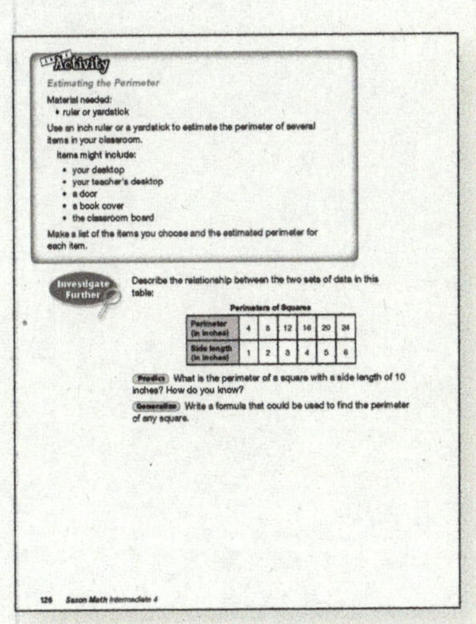

Activity

Estimating the Perimeter

Material needed:

- ruler or yardstick

Use an inch ruler or a yardstick to estimate the perimeter of several items in your classroom.

Items might include:

- your desktop
- your teacher's desktop
- a door
- a book cover
- the classroom board

Make a list of the items you choose and the estimated perimeter for each item.

Investigate Further

Describe the relationship between the two sets of data in this table:

Perimeters of Squares

Perimeter (in inches)	4	8	12	16	20	24
Side length (in inches)	1	2	3	4	5	6

Predict What is the perimeter of a square with a side length of 10 inches? How do you know?

Generalize Write a formula that could be used to find the perimeter of any square.

126 *Saxon Math Intermediate 4*

Concepts covered in the investigations will appear in subsequent Written Practices and will be assessed; therefore, investigations may not be skipped.

Using Math Launch, Grade 3

Math Launch provides teachers with a set of bulletin board posters and activity materials based on themes which are built into the ***Saxon Math*** *Intermediate 3* textbook. The bulletin board poster and accompanying Daily Math Worksheets serve as a precursor activity for the day's math lesson. The purpose of **Math Launch** is to help "jump start" your students to think about the math that will be presented in the day's math lesson.

Math Launch contains 3 main components:

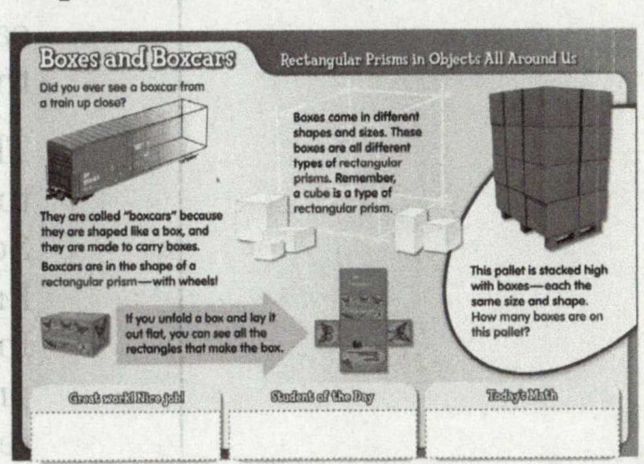

- 11 Thematic Bulletin Board Posters to accompany each of the 11 sections of the ***Saxon Math*** *Intermediate 3* textbook
- Daily Math Worksheets for each day of instruction
- The *Teacher Resource/Additional Activities* booklet that includes teacher notes for the worksheets

Using Technology

The Saxon Math program recognizes that technology is an important tool of instruction.

The program has been expanded to include several CDs for teachers and students.

- The ***Resources and Planner CD*** provides resources and helps the teacher plan the year's lessons.
- The ***Monitoring Student Progress eGradebook CD*** helps teachers track the progress of their students on Cumulative and Benchmark Tests. It also generates a variety of reports, including standards reports.
- The ***Test and Practice Generator CD*** allows teachers to create unlimited practice and test items in multiple formats.
- The ***Instructional Presentations CD*** provides ready-to-use presentations that are aligned to each lesson for instant teacher support.
- The ***Student Edition eBook CD*** automatically strengthens the school to home connection with convenient access to complete, easy to use texts at home.
- The ***Teacher's Manual eBook CD*** is the complete two-volume Teacher's Manual provided on CD so teachers can take advantage of technology in the classroom or elsewhere.
- The ***Answer Key CD*** has answers to Power-Up, Mental Math, Problem Solving, Lesson Practice, Written Practice, Early Finishers, and Investigation exercises. The pages provided on this CD are identical to the reduced student pages located in the Teacher's Manual.

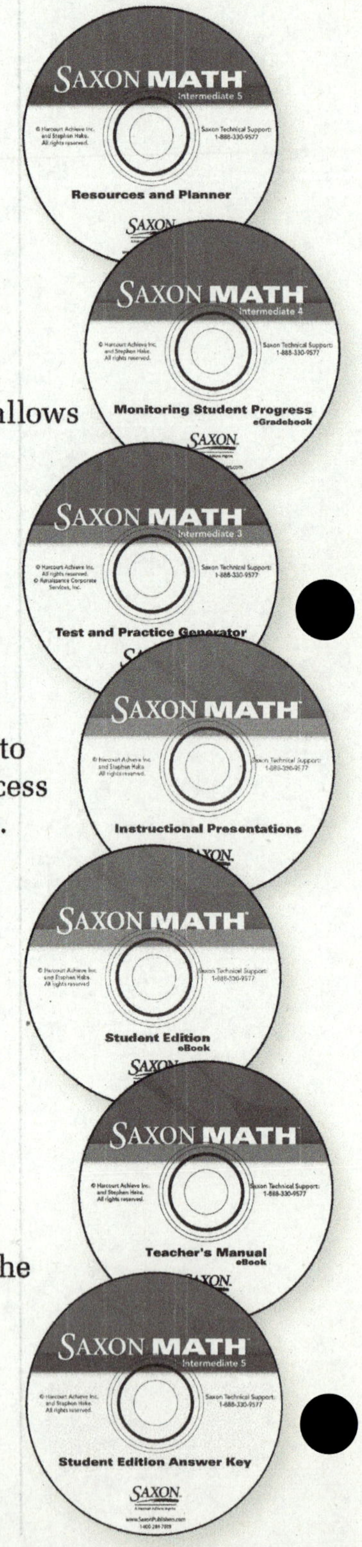

Because classrooms across the United States vary greatly in what technology is available, the Saxon Math program has also focused on a widely accessible tool—the Internet.

Additional information and resources can be found on the Web at www.SaxonMath.com, which provides an overview of each Saxon Math program. Free student activities are available online, including:

- **Calculator Activities**
  These calculator activities provide students with the opportunity to incorporate language and reasoning with technology. Examples and step-by-step instructions for using the calculator to solve problems are provided with each activity. There are problems included in the activities that are relevant to textbook lessons.

- **Interactive Online Activities**
  Making patterns that have symmetry, making arrays to show multiplication facts, and solving four different types of math story problems are examples of the variety of concepts that are covered in the interactive online activities. Students are also asked to answer questions about the activities that they complete.

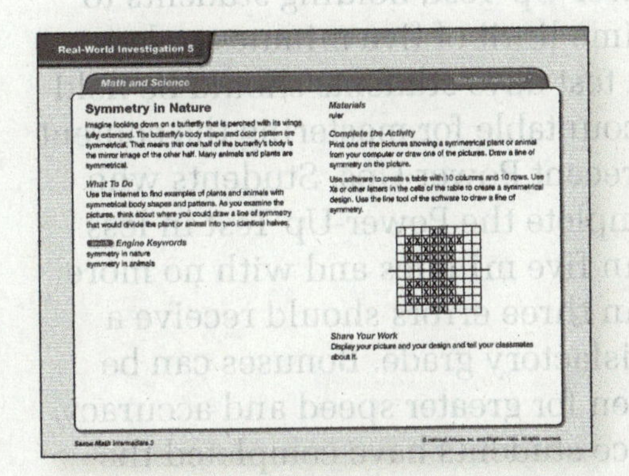

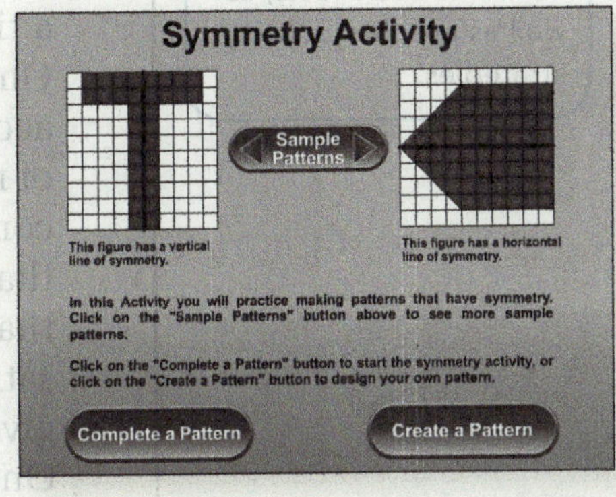

- **Real-World Investigations**
  These enrichment exercises challenge students to apply their mathematical knowledge as they complete each investigation for their grade level.

Tip

The calculator icon that appears next to the New Concept in some lessons will indicate that an online calculator activity is available that corresponds to that concept.

Tip

An online icon placed next to the New Concept in some lessons indicates that a student activity that corresponds to the skill being presented in the New Concept is available online.

Use these online activities to provide students with additional practice, remediation, or enrichment. Students enjoy learning and practicing mathematical concepts online. These online activities also provide a way for parents to work with their children on math at home. The online activities are provided in both English and Spanish.

Assessing Student Progress

Students' acquisition and retention of skills and concepts is evaluated through built-in, frequent cumulative assessments. A Power-Up Test and a Cumulative Test should be given after every fifth lesson beginning after Lesson 10. These assessments cover a wide variety of concepts that have been previously taught and practiced.

Three activities are scheduled for each test day:

1. Power-Up Test
2. Cumulative Assessment
3. Test-Day Activity or a Performance Task

Because the activities on a test day are likely to fill the class period, we recommend that the previous day's homework assignment be reviewed on the day following the test.

Tip

Use the Power-Up Tests to assess basic facts, as well as problem solving strategies.

Power-Up Tests

Begin the test day with the specified Power-Up Test, holding students to a time limit of five minutes or less. On test days students should be held accountable for mastering the content of recent Power Ups. Students who complete the Power-Up Test in less than five minutes and with no more than three errors should receive a satisfactory grade. Bonuses can be given for greater speed and accuracy. Once students have completed the Power-Up Test, begin the Cumulative Test.

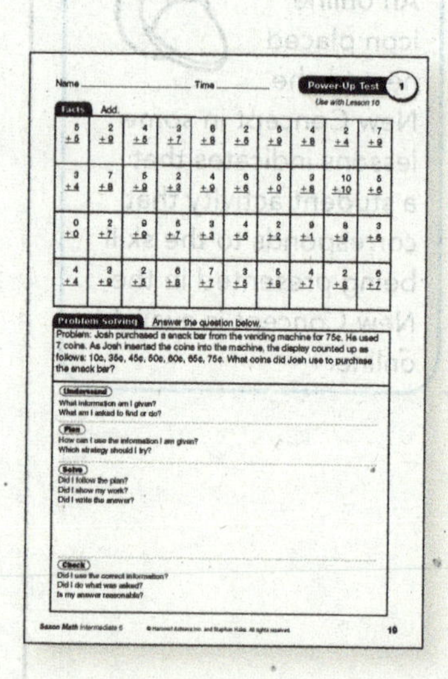

Name ________ Time ________

Power-Up Test 1

Use with Lesson 10

Facts Add.

5 + 5	2 + 9	4 + 5	3 + 7	6 + 8	2 + 5	6 + 9	4 + 8	2 + 4	7 + 9
3 + 4	7 + 8	5 + 9	2 + 3	4 + 9	6 + 6	5 + 0	9 + 8	10 + 10	5 + 6
0 + 0	2 + 7	9 + 9	6 + 7	3 + 3	4 + 5	2 + 2	9 + 1	8 + 9	3 + 6
4 + 4	3 + 9	2 + 5	6 + 8	7 + 7	3 + 5	5 + 8	4 + 7	2 + 6	6 + 7

Problem Solving Answer the question below.

Problem: Josh purchased a snack bar from the vending machine for 75¢. He used 7 coins. As Josh inserted the coins into the machine, the display counted up as follows: 10¢, 35¢, 45¢, 50¢, 60¢, 65¢, 75¢. What coins did Josh use to purchase the snack bar?

Understand
What information am I given?
What am I asked to find or do?

Plan
How can I use the information I am given?
Which strategy should I try?

Solve
Did I follow the plan?
Did I show my work?
Did I write the answer?

Check
Did I use the correct information?
Did I do what was asked?
Is my answer reasonable?

Saxon Math Intermediate 5 © Harcourt Achieve Inc. and Stephen Hake. All rights reserved. 10

Cumulative Tests

> **Tip**
>
> Each test has twenty problems. Each problem is worth five points for a total value of 100 points. Be sure to enforce proper labeling. Deduct one or two points from each answer with an incorrect label.

Cumulative Tests are designed to reward students and to provide teachers with diagnostic information. The Cumulative Test allows students to demonstrate the skills they have developed, and it builds confidence that will benefit students when they encounter comprehensive standardized tests.

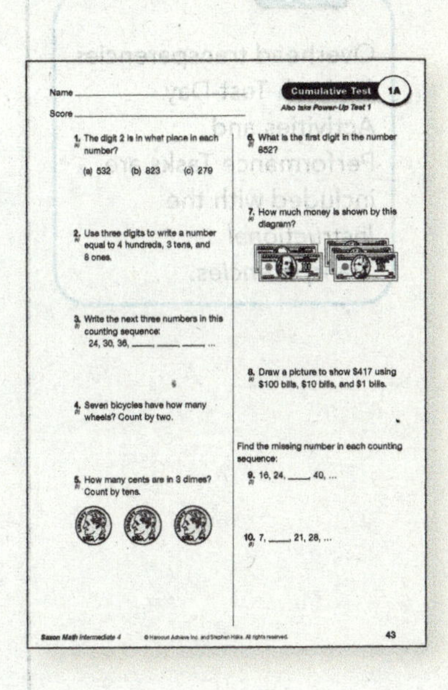

Name ______________________

Score ______________________

Cumulative Test 1A

Also take Power-Up Test 1

1. The digit 2 is in what place in each number?
   (a) 532 (b) 823 (c) 279
2. Use three digits to write a number equal to 4 hundreds, 3 tens, and 8 ones.
3. Write the next three numbers in this counting sequence:
   24, 30, 36, ____, ____, ____, ...
4. Seven bicycles have how many wheels? Count by two.
5. How many cents are in 3 dimes? Count by tens.
6. What is the first digit in the number 852?
7. How much money is shown by this diagram?
8. Draw a picture to show $417 using $100 bills, $10 bills, and $1 bills.

Find the missing number in each counting sequence:

9. 16, 24, ____, 40, ...
10. 7, ____, 21, 28, ...

Saxon Math Intermediate 4 © Harcourt Achieve Inc. and Stephen Hake. All rights reserved. 43

Administering the Cumulative Tests according to the schedule is essential. The Cumulative Tests are available after every five lessons, begining after Lesson 10. Following the schedule allows students sufficient practice on new topics before they are assessed on those topics.

It is important for students to show their work for each problem on the Cumulative Test. There are three different answer forms, Answer Forms A, B, or C, located in the *Assessment Guide* that provide a space for students to show their work for each problem. Decide which form you want to use and make a copy for each student to use for the Cumulative Tests.

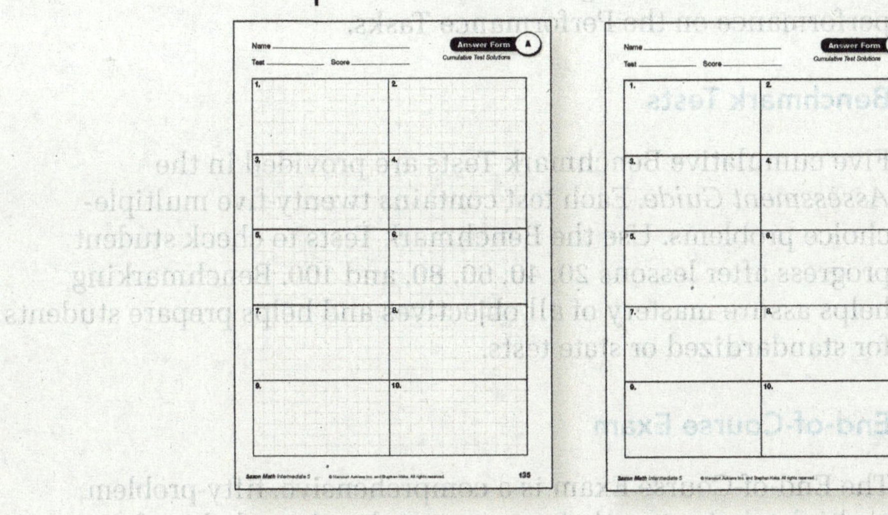

Name ______________________

Test ________ Score ________

Answer Form A

Cumulative Test Solutions

1. 2. 3. 4. 5. 6. 7. 8. 9. 10.

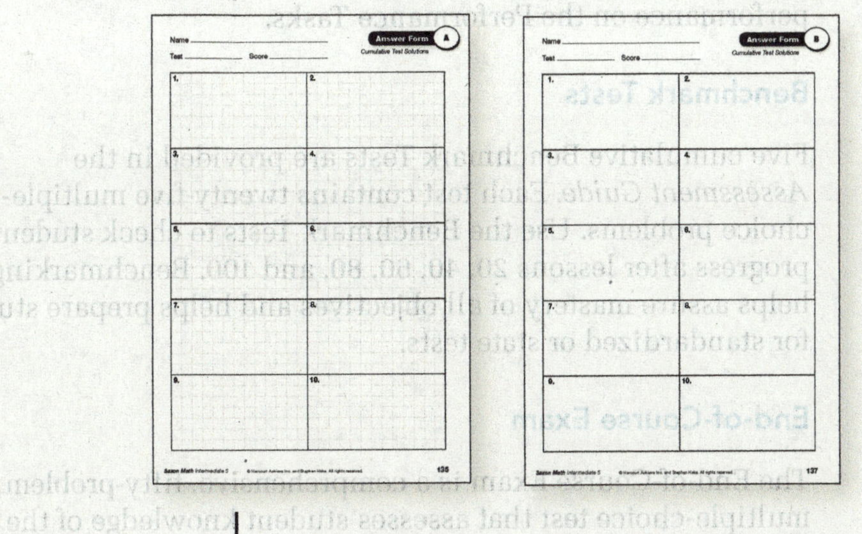

Name ______________________

Test ________ Score ________

Answer Form B

Cumulative Test Solutions

1. 2. 3. 4. 5. 6. 7. 8. 9. 10.

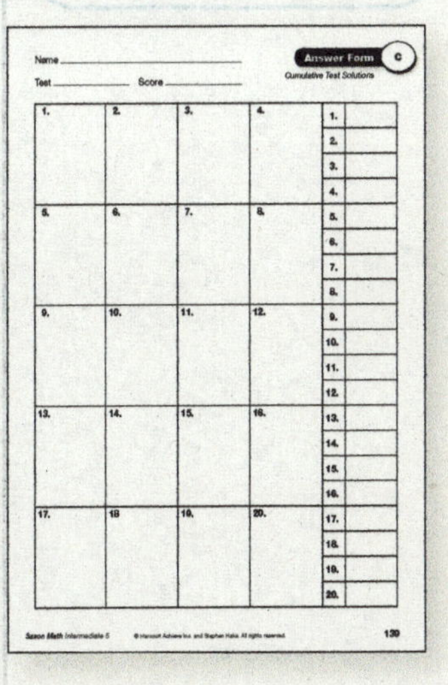

Name ______________________

Test ________ Score ________

Answer Form C

Cumulative Test Solutions

1. 2. 3. 4. 5. 6. 7. 8. 9. 10. 11. 12. 13. 14. 15. 16. 17. 18. 19. 20.

The answers for the Cumulative Tests are found in the Assessment Guide.

Tip

Overhead transparencies for both Test-Day Activities and Performance Tasks are included with the *Instructional Transparencies.*

Test Day Activities and Performance Tasks

After students finish their tests, move on to the Test-Day Activity or to the Performance Task. These activities or tasks use time available after the Power-Up Test and Cumulative Test to provide an enriching learning experience for students. Most activities or tasks can be completed in 10–15 minutes. Instructions and worksheets for the Test-Day Activities are in the *Instructional Masters* booklet. Instructions and worksheets for the Performance Tasks are in the *Performance Tasks* booklet.

The Performance Tasks allow students to explore topics in the real world and to explain their thinking with open-ended questions. Rubrics are included to assist you with the evaluation and scoring of student performance on the Performance Tasks.

Tip

The Performance Task Scoring Rubrics and an answer key are provided in the Performance Tasks booklet.

Benchmark Tests

Five cumulative Benchmark Tests are provided in the *Assessment Guide.* Each test contains twenty-five multiple-choice problems. Use the Benchmark Tests to check student progress after lessons 20, 40, 60, 80, and 100. Benchmarking helps assure mastery of all objectives and helps prepare students for standardized or state tests.

End-of-Course Exam

The End-of-Course Exam is a comprehensive, fifty-problem, multiple-choice test that assesses student knowledge of the content presented in the course. The cumulative assessment should be given as late in the year as possible.

Monitoring Student Progress

Student progress needs to be monitored and a student who scores below 80 percent on an assessment needs immediate individual help. Without remediation, these students often fall hopelessly behind within a few weeks. The amount of help needed depends on the nature of the errors. Is the student making careless mistakes in computation, or is the student having difficulty understanding concepts and procedures? Saxon Math provides two methods for tracking student progress.

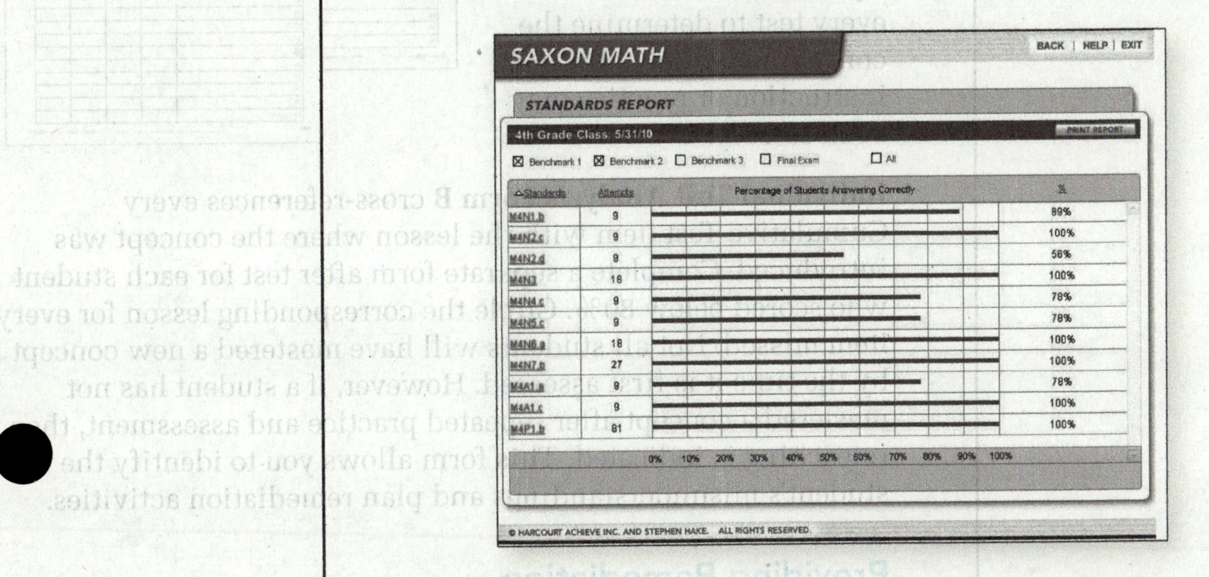

Student progress can be monitored using the ***Monitoring Student Progress: eGradebook.*** Enter students' scores on Cumulative Tests and Benchmark Tests into the *eGradebook*. Use the report titled Benchmark Standards Report to determine which objectives were assessed and the level of mastery for each student.

The Class Test-Analysis Form and the Individual Test-Analysis Form provided in the *Assessment Guide* can also be used to monitor student progress and to aid your analysis of each assessment.

Beginning with Cumulative Test 1, use **Class Test Analysis Form A** to record those test items that students have missed. By reviewing the column for items that appear repeatedly, you can quickly determine which test items are causing students the most trouble. Update this form after every test to determine the concept for which additional instruction or practice may be necessary.

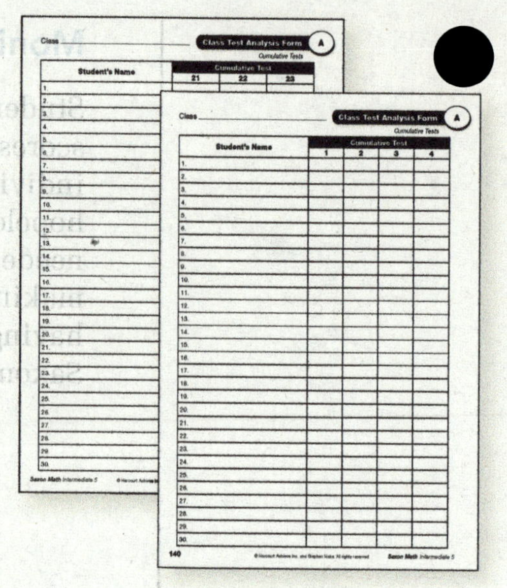

Class ______ Class Test Analysis Form A

Cumulative Tests

Student's Name	Cumulative Test		
	21	22	23
1.			
2.			
3.			
4.			
5.			
6.			
7.			
8.			
9.			
10.			
11.			
12.			
13.			
14.			
15.			
16.			
17.			
18.			
19.			
20.			
21.			
22.			
23.			
24.			
25.			
26.			
27.			
28.			
29.			
30.			

Saxon Math Intermediate 5

Class ______ Class Test Analysis Form A

Cumulative Tests

Student's Name	Cumulative Test			
	1	2	3	4
1.				
2.				
3.				
4.				
5.				
6.				
7.				
8.				
9.				
10.				
11.				
12.				
13.				
14.				
15.				
16.				
17.				
18.				
19.				
20.				
21.				
22.				
23.				
24.				
25.				
26.				
27.				
28.				
29.				
30.				

140 © Harcourt Achieve Inc. and Stephen Hake. All rights reserved. Saxon Math Intermediate 5

Individual Test Analysis Form B cross-references every Cumulative Test item with the lesson where the concept was introduced. Complete a separate form after test for each student who scored below 80%. Circle the corresponding lesson for every item missed. Not all students will have mastered a new concept by the time it is first assessed. However, if a student has not mastered a concept after repeated practice and assessment, then reteaching is indicated. This form allows you to identify the student's misunderstandings and plan remediation activities.

Providing Remediation

Saxon's methodology of continual, distributed practice provides students and teachers with a built-in remediation feature. Saxon's distributed approach provides students with numerous opportunities to practice skills over a longer period of time, thus giving them the time needed to master a concept. Occasionally, however, students might still need additional remediation, so the following options are available to teachers:

- Review the lesson again or assign the **Reteaching Master**. There is a Reteaching Master available for each lesson.

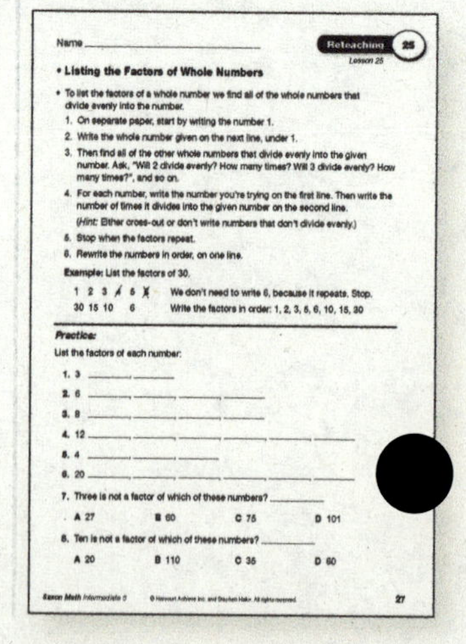

Name ______ Reteaching 25

Lesson 25

• **Listing the Factors of Whole Numbers**

• To list the factors of a whole number we find all of the whole numbers that divide evenly into the number.

1. On separate paper, start by writing the number 1.
2. Write the whole number given on the next line, under 1.
3. Then find all of the other whole numbers that divide evenly into the given number. Ask, "Will 2 divide evenly? How many times? Will 3 divide evenly? How many times?", and so on.
4. For each number, write the number you're trying on the first line. Then write the number of times it divides into the given number on the second line.

(*Hint:* Either cross-out or don't write numbers that don't divide evenly.)

5. Stop when the factors repeat.
6. Rewrite the numbers in order, on one line.

Example: List the factors of 30.

1 2 3 4 5 6 — We don't need to write 6, because it repeats. Stop.

30 15 10 6 — Write the factors in order: 1, 2, 3, 5, 6, 10, 15, 30

Practice:

List the factors of each number:

1. 3 ______, ______
2. 6 ______, ______, ______, ______
3. 8 ______, ______, ______, ______
4. 12 ______, ______, ______, ______, ______, ______
5. 4 ______, ______, ______
6. 20 ______, ______, ______, ______, ______, ______
7. Three is not a factor of which of these numbers? ______

A 27 B 60 C 75 D 101

8. Ten is not a factor of which of these numbers? ______

A 20 B 110 C 35 D 60

Saxon Math Intermediate 5 © Harcourt Achieve Inc. and Stephen Hake. All rights reserved. 27

Each Reteaching Master contains a concise summary of the New Concept introduced in a lesson or investigation, including annotated examples. The lesson summary is followed by a set of practice items to support reteaching and remediation of the New Concept.

- Use the ***Test and Practice Generator CD*** to design and create a worksheet for remediation purposes.

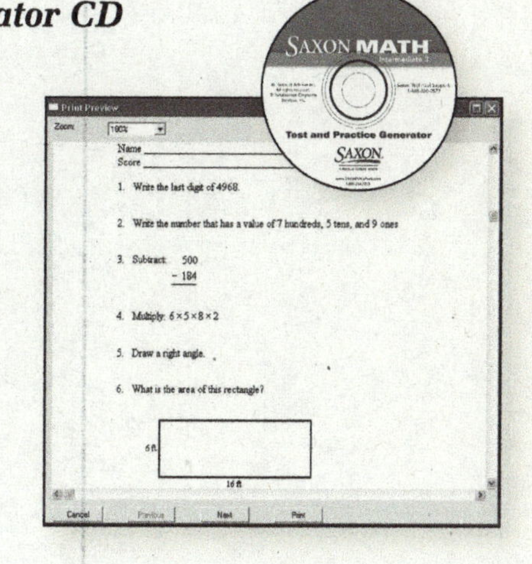

If you decide to use the *Test and Practice Generator* software to design worksheets for remediation purposes, create only a few problems at a time. During class time designated for working on the Written Practice, pull one or more students aside to work on a targeted concept. Complete one or two problems as a group; then have students work on one or two more problems independently. Repeat this procedure over a period of one or two weeks. This method of remediation, when used in conjunction with the distributed practice, will help students achieve conceptual understanding of the problem area.

> **REMEMBER!** Problems generated with the *Test and Practice Generator* do not replace the Written Practice. Students should continue working at the pace of one Written Practice a day with the exception of test and investigation days.

Within each classroom there are students with special needs, such as English learners, struggling or at-risk students, and advanced learners. Saxon Math has built-in support for all students. Also included in the program are support features that will help teachers differentiate and customize instruction to meet the needs of all students.

Meeting the Needs of English Learners

Throughout the student text, ESL/ELL students will find structures to help them acquire mathematical understanding. These structures include visual models, hands-on activities, and Math Language prompts.

Additionally, the English Learners teacher notes focus on language acquisition, not on reteaching or simplifying the math. These notes are based on a proven approach which uses the following steps:

1. Define/Hear
2. Discuss/Explain
3. Apply/Use
4. Model/Connect

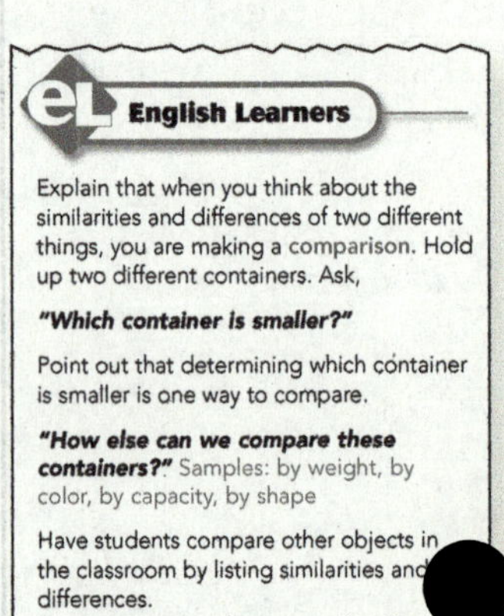
English Learners

Explain that when you think about the similarities and differences of two different things, you are making a comparison. Hold up two different containers. Ask,

"Which container is smaller?"

Point out that determining which container is smaller is one way to compare.

"How else can we compare these containers?" Samples: by weight, by color, by capacity, by shape

Have students compare other objects in the classroom by listing similarities and differences.

Teachers will also find the *English Learners Handbook* to be a valuable resource as they strive to meet the needs of English learners in the mathematics classroom. This book covers:

- Techniques for making instruction comprehensible to all students
- Characteristics of a mathematics classroom environment that is supportive to English learning
- Mathematics vocabulary in nine languages
- Prefixes and suffixes that serve similar functions in English and Spanish
- Spanish cognates for mathematics vocabulary and tips for teaching with cognates
- Background information for eleven primary languages
- Grammar and phonics transfer issues for different languages

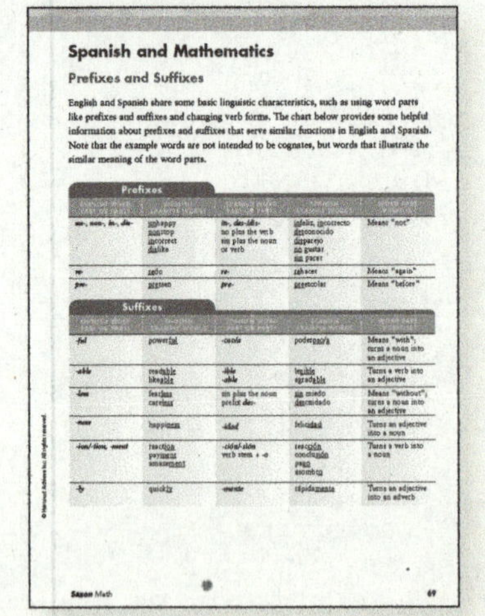
Spanish and Mathematics

Prefixes and Suffixes

English and Spanish share some basic linguistic characteristics, such as using word parts like prefixes and suffixes and changing verb forms. The chart below provides some helpful information about prefixes and suffixes that serve similar functions in English and Spanish. Note that the example words are not intended to be cognates, but words that illustrate the similar meaning of the word parts.

Prefixes

[illegible]	[illegible]	[illegible]	[illegible]	[illegible]
un-, non-, in-, dis-	unhappy nonstop incorrect dislike	in-, des-/dis- no plus the verb sin plus the noun or verb	infeliz, incorrecto desconocido disparejo no gustar sin parar	Means "not"
re-	redo	re-	rehacer	Means "again"
pre-	preteen	pre-	preescolar	Means "before"

Suffixes

[illegible]	[illegible]	[illegible]	[illegible]	[illegible]
-ful	powerful	-oso/a	poderoso/a	Means "with"; turns a noun into an adjective
-able	readable likeable	-ible -able	legible agradable	Turns a verb into an adjective
-less	fearless careless	sin plus the noun prefix des-	sin miedo descuidado	Means "without"; turns a noun into an adjective
-ness	happiness	-idad	felicidad	Turns an adjective into a noun
-ion/-tion, -ment	reaction payment amazement	-ción/-sión verb stem + -o	reacción conclusión pago asombro	Turns a verb into a noun
-ly	quickly	-mente	rápidamente	Turns an adjective into an adverb

Saxon Math 69

Tip

Also available is the *English Learners Handbook*, a valuable tool to help teachers respond to the needs of students learning English.

For Spanish speakers, the Glossary in the student text provides a Spanish translation and a Spanish definition of each math term. The complete program is also available in Spanish.

Meeting the Needs of Struggling Learners

Saxon Math includes a number of support features in the Teacher's Manual to help various categories of struggling learners.

- **Inclusion** helps teachers to accommodate special needs in the classroom so that all students have an opportunity to grasp new concepts.
- **Alternative Approach: Using Manipulatives** suggest a variation to the New Concept presentation in a lesson that might help students gain understanding. This feature gives an alternative for concept development through the use of hands-on manipulative activities.
- The **Errors and Misconceptions** section alerts the teacher to common student misunderstandings that may block proper development of the concept.
- The **Math Language** box on the Lesson Planning and Preparation page gives a list of math vocabulary that will be introduced and reviewed in that lesson.
- **Flexible Grouping** includes ideas for various grouping strategies tied to specific lessons.

Alternative Approach: Using Manipulatives

To help students understand the concept of comparing and ordering decimal numbers, have students model each number being compared using money manipulatives. Remind students that pennies are used to represent hundredths and dimes are used to represent tenths. Have students read each decimal number as a money amount when comparing or ordering.

The consistent lesson format of Saxon Math provides a predictable routine that enables all learners to be successful.

Meeting the Needs of Advanced Learners

Saxon Math also provides advanced learners with many opportunities for expanding their concept development.

- **Early Finishers** in Written Practice offers the opportunity to deepen mathematical learning with problem-solving, cross curricular, and enrichment activities.
- **Investigate Further** in the Investigations allows students to expand their knowledge of the investigation concepts, sharpen their higher order thinking skills, and explore more connections.
- **Extend the Example** and **Extend the Problem** suggestions in the Teacher's Manual provide opportunities to broaden concept development by expanding on an example or a problem.

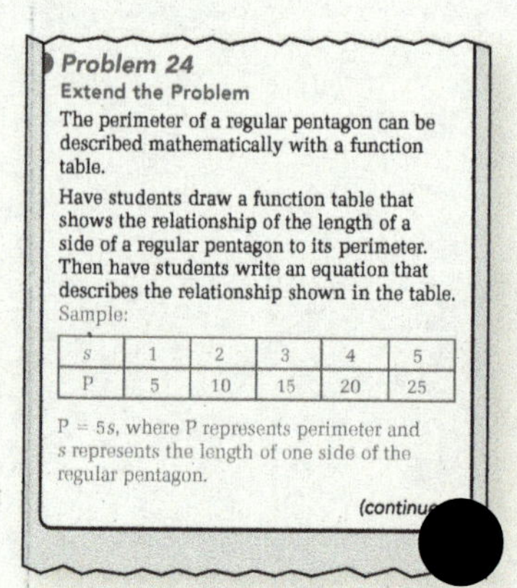

Problem 24

Extend the Problem

The perimeter of a regular pentagon can be described mathematically with a function table.

Have students draw a function table that shows the relationship of the length of a side of a regular pentagon to its perimeter. Then have students write an equation that describes the relationship shown in the table.

Sample:

s	1	2	3	4	5
P	5	10	15	20	25

$P = 5s$, where P represents perimeter and s represents the length of one side of the regular pentagon.

(continu

- The **Online Activities** include real-world investigations, calculator activities, and exploration activities. These are available on the Web at www.SaxonMath.com. Advanced learners who are working ahead of the class can use these activities.

Adaptations for Saxon Math

Adaptations for Saxon Math is an intervention program that addresses the needs of students who struggle with math. Meeting the needs of struggling math students is challenging, but *Adaptations* materials help teachers achieve that goal.

Adaptations materials are designed to ensure success for a range of students, including students who have difficulties with the following:

- visual-motor integration
- distractibility or lack of focus
- receptive language
- number reversal in reading and copy work
- fine motor coordination
- math anxiety
- verbal explanation
- spatial organization

Tip

The carefully structured layout of the Written Practice pages helps struggling students focus on mastering the concept, rather than figuring out the directions.

Adaptations materials provide alternate teaching strategies, modifications of lessons and tests, student reference materials, and extra practice on select topics. They can be used in a variety of classroom settings, including inclusion and pullout programs. For more information on how to use the materials according to the needs of your classroom, see the *Adaptations for Saxon Math Teaching Guide.*

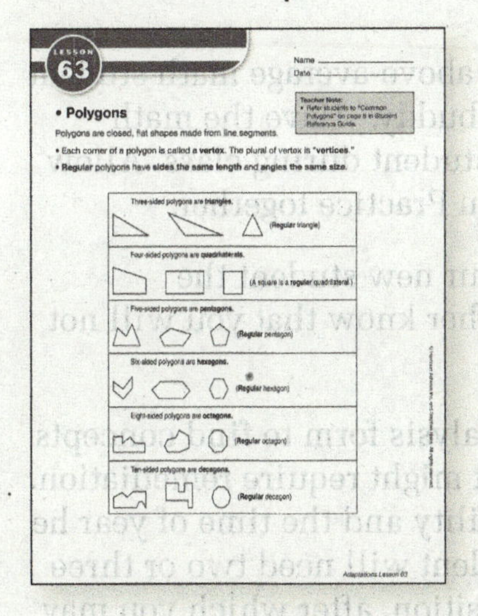

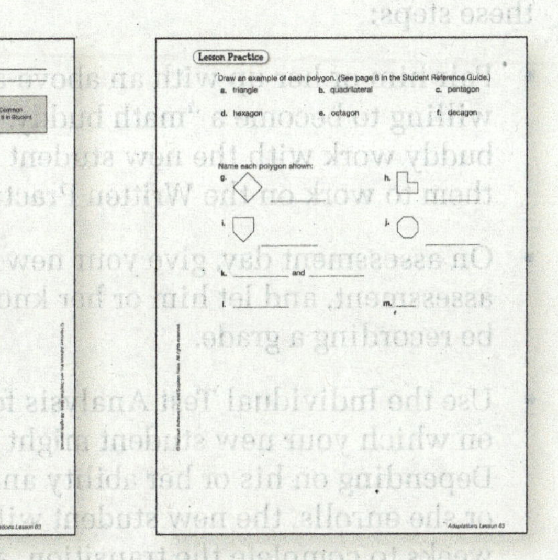

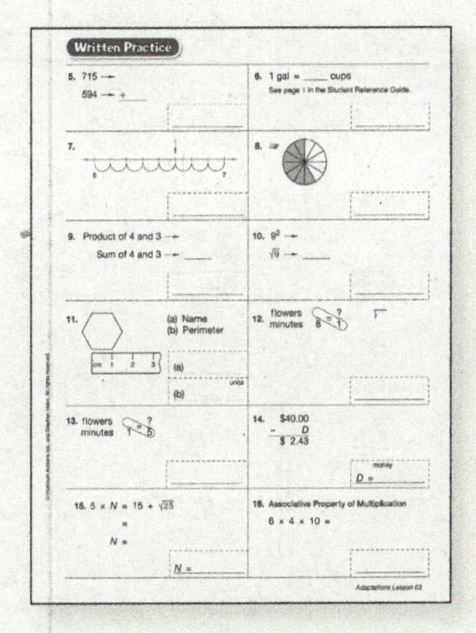

Transitioning into Saxon Math

Tip

Don't worry if the first ten to fifteen lessons seem easy. Early lessons are designed to review concepts from the previous year and to build a foundation that will make it easy for students, to learn program procedures.

If this is your class's first year using Saxon Math, it is important for you to establish your procedures and expectations during the first week of the school year.

Your Class

During the first week, demonstrate to your students how you want them to take notes. Show students the process for completing the daily Written Practice. Don't forget to model how to show work for solving each problem, and stress the importance of labeling each answer. Have students work along with you as you show the steps needed to solve each problem. After the first week students should have a firm understanding of classroom procedures and expectations.

New Students

Teachers often worry that new students will have a difficult time transitioning into Saxon Math because of the unique nature of the program. This is rarely the case. With assistance, students moving into your class after the start of the school year will easily and successfully transition into Saxon Math.

When a non-Saxon student first joins your classroom, follow these steps:

- Pair him or her up with an above-average math student willing to become a "math buddy." Have the math buddy work with the new student during class. Allow them to work on the Written Practice together.
- On assessment day, give your new student the assessment, and let him or her know that you will not be recording a grade.
- Use the Individual Test Analysis form to find concepts on which your new student might require remediation. Depending on his or her ability and the time of year he or she enrolls, the new student will need two or three weeks to complete the transition, after which you may begin recording a grade for his or her assessments.

Involving Parents

Parent involvement greatly improves a student's chance for success. It is important to explain to parents early in the school year how they can help their child be successful in math. This can be done by sending the provided parent letter home with students and by explaining the Saxon Math program at the first open house of the year. To ensure that the parents receive the letter, you may require the students to return the letters bearing the parents' signatures.

Tip

A letter to parents explaining the Saxon Math program and student success is available in English and Spanish. It can be found in the Instructional Masters booklet.

Dear Parent or Guardian:

This year your child will be using a ***Saxon Math*** textbook. ***Saxon*** books are unlike traditional texts in which students are taught and expected to learn an entire mathematical concept in one day. Instead of practicing 30 problems of the same kind, ***Saxon Math*** builds mastery through daily practice of skills and concepts taught in earlier lessons.

Saxon textbooks divide concepts into small, easily grasped pieces called increments. A new increment is presented each day, and students work only a few problems each day involving the new material. The remaining homework problems cover previously introduced concepts. Thus, every assignment (and every test) is a review of material covered up to that point, providing students with the time and opportunity to understand and remember necessary concepts.

This is where your involvement is crucial. *It is essential that your child complete each day's assignment.* You can expect your child to complete three to five lessons per week with a test after every fifth lesson. Please ask to see your child's daily assignment, and encourage him or her to complete all the work. Students who complete all the daily assignments will have much to be proud of at test time, so you will have much to praise.

Some key words in the preface of your child's textbook are printed in boldface. They remind your child of the importance of the daily exercises. Those words are worth repeating here:

> *Solving these problems day after day is the secret to success. Work every problem in each Lesson Practice, Written Practice, and investigation. Do your best work and you will experience success and true learning, which will stay with you and serve you well in the future.*

Learning mathematics is similar to learning a foreign language, a musical instrument, or an athletic skill. Long-term practice is necessary to develop and maintain mastery of the problem-solving skills needed for success in math-intensive disciplines.

Your child's experience in mathematics will be a direct result of his or her attitude and effort. A positive attitude and a commitment to long-term practice greatly improve the chances of your child progressing into higher-level mathematics courses and experiencing success in other fields that require mathematical understanding. Through our combined efforts your child can experience success at this level and learn the foundational skills needed for the next level.

Sincerely,

At the first open house, it is helpful to explain to parents that, although the program might appear easy at first, the early lessons review and solidify foundational skills, and will become steadily more challenging as the year progresses. You may show parents a few assessments from the end of the year to illustrate this.

Consider planning a Saxon Math Night during which students and parents attend a one-hour class in the students' classroom. During the hour, students can demonstrate the daily Power-Up, present a lesson that they helped choose, and show how they complete the daily Written Practice. This will give parents a firsthand understanding of the Saxon Math program.

Preparing for Standardized or State Testing

Tip

The Benchmark Tests provide practice with multiple-choice test items. Familiarity with this format will lead to success on standardized assessment tests.

Most teachers are required to administer a state test each year. State tests vary widely, with each state's test reflecting its specific learning objectives. Test items may be multiple-choice, open-ended, free response, gridded response, or a combination of two or three of these. The dates on which tests are given also vary, with some states testing in the fall, some in early spring, and others in late spring.

Here are some time-efficient ways to provide practice for your students and ensure that the tested objectives are covered before the testing date.

- Familiarize yourself with your state test's objectives and format. Ask your school principal, staff developer, or math coordinator for the list of objectives covered on the test and for released sample items. If sample responses are provided for open-ended questions, study them carefully to identify the important characteristics of acceptable answers.
- Also, ask your administrator for the last available test results for your students. Study their strengths and weaknesses to plan for your class.
- To ensure that the objectives on the test are taught prior to your testing date, use the pacing calendar you created at the beginning of the school year. Lesson mapping will allow you to identify how many lessons you will teach before the test. Generally, each day you will teach one lesson or investigation or give an assessment. If you test in March, make sure that you teach one lesson a day until the test is given. If your testing date is in late April or May, you may be able to use some days for extra practice or review throughout the year. Using your pacing calendar will allow you to plan for proper test preparation.
- Use the **Test-Taking Strategies Guide** to provide guidance for taking standardized test. This book reviews vocabulary, process skills, and figures that may be used on a standardized test. Examples and strategies are given to help students solve different types of problems. To give students the opportunity to practice the skills presented in the book, strategy practice problems are provided with each example. Also, a practice test is located at the end of the *Test-Taking Strategies Guide*. This practice test can be used to familiarize your students with test-taking formats and procedures.

Manipulative Kits

The manipulative kit is a set of hands-on items that students can use to explore the mathematical concepts presented in the Student Edition and the Teacher's Manual. There are two kits available.

The **Basic Manipulative Kit** includes the following items:

- Balance Scale with Geometric Weights
- Base 10 Cube
- Base 10 Flats
- Base 10 Rods
- Base 10 Units
- Bills, $1
- Bills, $100
- Bills, $5
- Clock, Large Demonstration
- Clock, Student
- Coin Set
- Color Tiles
- Compass, Safety
- Counters, Two-Color
- Cups, Measuring Dual
- Dice (Dot)
- Fraction Circles
- Geoboards
- Meterstick, Folding
- Number Line, Teachers
- Protractors
- Relational Geosolids
- Rubber Bands
- Ruler
- Spinners
- Tape Measure
- Thermometer, Classroom
- Wooden Color Cubes

The **Overhead Manipulative Kit** is available as an optional purchase. The following items are included in this kit:

- Base 10 Overhead Set
- Color Tiles Overhead
- Compass, Safety Overhead
- Counters Overhead, Two-Color
- Fraction Circles, Overhead, Basic
- Geoboards Overhead
- Pattern Blocks Overhead
- Spinners Overhead, Circle